TECHNIQUES AND EXPERIMENTS FOR ADVANCED ORGANIC LABORATORY

CHARLES M. GARNER
Baylor University

JOHN WILEY & SONS, INC.
New York • Chichester • Brisbane • Toronto • Singapore • Weinheim

The Author of this manual has outlined extensive safety precautions in each experiment. Ultimately, it is your responsibility to practice safe, laboratory guidelines. The Author and Publisher disclaim any liability for any loss or damage claimed to have resulted from, or related to, the experiments.

ISBN 0-471-17045-3

Printed in the United States of America

10 9 8 7 6 5 4 3 2 1

Printed and bound by Bradford & Bigelow, Inc.

Preface

The transition from the Sophomore organic lab to doing research is complicated by several factors - differences in the size of reactions, the use of more sophisticated and expensive equipment, exposure to difficult-to-handle materials, a heavier reliance on analytical techniques and instrumentation, and the need to learn to design procedures for new reactions. Because of this, a need for a research-oriented organic laboratory course has been recognized at many schools. However, I (and others) had been unable to identify a suitable laboratory manual for such a laboratory. This manual was written to introduce students to a variety of techniques which are used in research, including the most useful instrumental analyses (NMR, capillary GC and GC-MS). Indeed, several of the experiments are designed to illustrate the power of modern instrumentation, particularly capillary GC and NMR. I have also attempted to choose especially interesting experiments, several of which require "detective work" to solve, similar to what is encountered in research. Finally, I have suggested optional ideas (not tested procedures) for more in-depth and independent studies in the "Exploring Further ..." sections. This text should be of use in Junior- or Senior-level advanced organic laboratories, as well as in the second semester of Sophomore organic laboratories where the instructor wishes to provide an especially rigorous laboratory experience.

The material in this text generally assumes an understanding of Sophomore-level organic laboratory techniques. It is intended that this laboratory manual be used in conjunction with a good spectroscopy text which gives a practical coverage of IR, proton and carbon NMR, and mass spectroscopy. In addition, we (and others) have used Zubrick's "The Organic Chemistry Laboratory Survival Manual" (John Wiley and Sons, 1992), which is a detailed and humorous discussion of laboratory pitfalls that is suitable for either a Sophomore or an advanced laboratory student.

Many participated in various aspects of this project. I am grateful to several at Wiley: Sharon Nobel and Joan Kalkut for early encouragement, and Jennifer Yee, Bonnie Cabot, and Nedah Rose for carrying the project to completion. Proofreading by Elisabeth Belfer greatly improved the consistency of the format. I wish to thank Dr. Robert Walkup of Texas Tech University for providing much of the Diels-Alder experiment and also for thorough reviews of the manuscript. I appreciate the initial encouraging reviews given by Kraig Steffan of Fairfield University and James Hansen of Seton Hall University. Most of the experiments were developed during four semesters (1993-96) of Chemistry 4237 (Advanced Organic Laboratory), a Junior/Senior-level laboratory course at Baylor University. Baylor students Farima Farzaneh and especially the wonderful Lisa Alvarez helped work out experiments and collect spectra. Sunil Aggarwal worked out the procedure for the prediction of GC retention times. Finally, without the facilities provided by Baylor University and the Department of Chemistry, and their commitment to excellence in undergraduate education, this project could not have been accomplished. I would welcome any questions, comments or suggestions.

Charles M. Garner
October, 1996

About the author: Charles M. Garner is an Associate Professor of Chemistry at Baylor University. He received his B.S. from the University of Nevada, Las Vegas, and his Ph.D. from the University of Colorado. After working for a time in industry, he held an NIH postdoctoral fellowship at the University of Utah. In addition to conducting research in organic and organometallic chemistry, Professor Garner enjoys developing new experiments and applications of chemistry instrumentation in teaching laboratories. He has directed organic laboratories at both the Sophomore and advanced levels for several years.

Table of Contents

INTRODUCTION

PREPARATION AND ISOLATION OF PRODUCTS

PURIFICATION AND ANALYSIS TECHNIQUES

vi

EXPERIMENTAL PROCEDURES

INTRODUCTION

Purpose of the Text

This text is intended as an introduction to the most useful techniques encountered in organic chemistry research. This includes experience in handling very reactive materials, accomplishing difficult chromatographic separations, and especially becoming familiar with the most powerful methods of instrumental analysis. The experiments are designed to illustrate the power of NMR, IR, GC-MS, and gas chromatography in a variety of contexts. Capillary gas chromatography allows rapid, high-resolution analyses, yet tends to be quite under-utilized in organic lab courses, so I have placed some emphasis on this technique. Also, several experiments illustrate advanced aspects of NMR spectroscopy, including multinuclear NMR. It is intended that a spectroscopy text which provides a moderately advanced (yet practical) treatment of NMR, IR, and MS be used in conjunction with this lab manual. The intended result is that, given the necessary knowledge and tools, you will develop a practical understanding of and an appreciation for research in organic chemistry.

Equipment

On the next page are shown the major pieces of "macroscale" glassware; learn each type and keep them clean. If one of your pieces should become cracked, replace it immediately. Never apply vacuum or pressure to fractured glass! Round bottom flasks are especially susceptible to "star" fractures: watch for them.

Laboratory Safety

Safety in chemistry laboratories is of the utmost importance. As in previous lab courses you may have taken, safety glasses or goggles are required at **all** times. You should avoid wearing shorts or other types of clothing which leave significant portions of skin exposed. Likewise, shoes should fully cover the feet (no sandals). When in the lab, do not put anything in your mouth. All volatile toxic materials should be handled only in the fume hoods. Wear gloves when there exists a risk of skin contact with toxic materials. Be careful not to expose organic liquids or vapors to spark or flame sources. Dispose of chemical wastes only in a manner authorized by your instructor, which will generally **not** be in the sink or wastebasket. You are not allowed to work alone in the laboratory. Unauthorized experiments are prohibited.

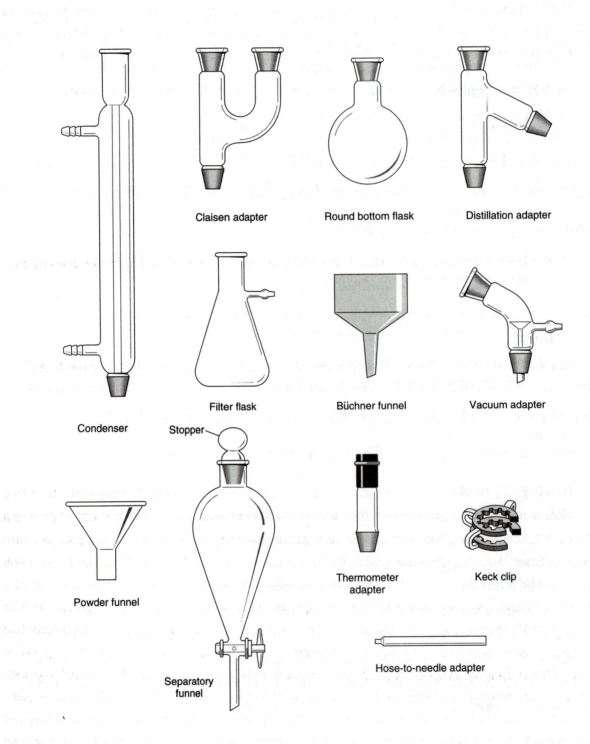

Figure 1: Common glassware and other items.

One aspect of advanced laboratory work is to develop an awareness of potential hazards various materials may present. The following terms are often used to describe particular hazards of various materials, and serve as an indication of how a given material should be handled. You should be familiar with these terms and their implications, and with common materials which have these properties. The label on a bottle may or may not note these hazards, but they should be given (when known) on the Material Safety Data Sheet (MSDS) provided by chemical suppliers.

Carcinogen: Causes cancer (e.g., methyl iodide, benzene, HMPA, chloroform, dioxane).

Deliquescent: Tending to absorb water from the air to the point of dissolving (e.g., $CaCl_2$, LiBr).

Hygroscopic: Absorbs water from the air (e.g., $MgSO_4$, DMSO, D_2O).

Lachrymator: Causes irritation and watering (tearing) of the eyes (e.g., benzyl bromide).

Mutagen: Causes mutations (e.g., 2-nitrobenzaldehyde).

Pyrophoric: Ignites spontaneously upon contact with air (e.g., t-butyllithium, trimethylaluminum, triethylborane, white phosphorus).

Sensitizer: Can cause serious allergic reactions on subsequent exposure (e.g., dicyclohexylcarbodiimide, iodomethane).

Sternutator: Causes sneezing and possibly lachrymation and vomiting (e.g., maleic anhydride, diphenylchloroarsine).

Teratogen: Causes birth defects (e.g., acrylamide, iodomethane, Thalidomide).

Vesicant: Causes blistering of skin (e.g., methyl iodide).

In addition, there are certain hazards particular to ethers. Diethyl ether and other low molecular weight ethers are extremely flammable. A more insidious hazard is that any ether with a CH bond next to the oxygen will slowly form peroxides upon exposure to air, which can cause two problems: (a) the peroxides can oxidize sensitive materials (like CuI), and (b) upon evaporation of the solvent, the higher-boiling peroxides concentrate and can detonate unexpectedly (an internal oxidation/reduction reaction). Diisopropyl ether is much worse that the other common ethers (diethyl ether and tetrahydrofuran). Prior to use, you should test any ethers which have had prolonged exposure to air (especially dry air) using commercially available starch/potassium iodide paper. This is done by evaporating a drop of the ether onto the paper followed by acidifying with 1 M HCl. An immediate change (within 1-3 seconds) to blue indicates the presence of peroxides. Slower color changes may be due to oxidation by atmospheric oxygen. Peroxide-contaminated ethers must be treated with an appropriate reducing agent followed by re-purification, or disposed of properly. The Aldrich catalog lists proper disposal procedures for the materials they sell.

Guidelines on Notebooks

Careful attention to thorough documentation of the laboratory notebook is extremely important in scientific training. Leave the first 2-3 pages of your notebook blank for a table of contents. Each experiment should always start on a new page. Always use a water-insoluble ink pen (**not** a felt-tip pen or a pencil) unless otherwise directed.

You should always:

• <u>Date</u> the experiment each day that you work in lab.

• <u>Draw out the reaction</u> that you hope will occur. You can also draw in possible undesired reactions if you wish.

• Note the <u>purpose</u> of the experiment.

• <u>Reference</u> the procedure you are following, if applicable.

• Provide a detailed <u>table of reagents</u>, which includes:

   - all the chemicals and solvents you will be using with a description of source and/or purity, if known (or appearance, if relevant).

   - the <u>pertinent</u> physical data for each compound (molecular weight, density for pure liquids, concentrations for solutions) required to calculate the mmoles used.

   - the actual amounts of materials you used, in the context you used them (if you weigh out a material, give the <u>weight</u> and not the volume that might correspond to).

   - the mmoles that each amount corresponds to (except for solvents).

   - the equivalents (i.e., mole ratios based on the limiting reagent) each amount corresponds to.

   - a <u>very</u> brief notation as to what hazards the material represents; this information may be obtained from the Aldrich catalog, or preferably from the material safety data sheets (MSDS) for that compound.

• Then describe what you did, including what size of flask you used, whether a stir bar was present, how you added the reagents and at what rate (if applicable), what temperatures were used, appearances during the experiment (especially color changes, homogeneity changes, gas or heat evolution, etc.). Be detailed enough that someone else could almost exactly reproduce your procedure using only your notebook. Date your notebook each new day you make an entry.

Note: Your instructor may or may not require you to use the third person when writing your notebook. The use of the first person is less formal, and in some cases may be more informative. For example, rather than writing "It was decided ...", something like "I decided ..." or "The instructor suggested ..." conveys more information. In more formal situations (for example, publications) the third person usually is used.

• All data from the characterization of products (such as product weight, yield, melting point, important IR frequencies, mass spectra peaks, and NMR chemical shifts and multiplicities, elemental analysis results, and GC retention times/area percents) should be entered into the notebook. *As much as possible, put a reduced copy of the spectrum or chromatogram into the notebook* (usually a 50 or 64% reduction of an appropriately sized original; we will work out good ways to do this for the various types of data). In addition to just *including* the data, you must also *interpret* it at an appropriate level. For example, what is it about your NMR spectrum or MS data that makes you think you have the correct compound? Always <u>draw in</u> pictures of important TLC plates (shade the spots with a <u>pencil</u>, but use ink for the rest; remember to note what solvent was used to develop the plate and what methods of visualization were used). In all reactions where you prepare and measure the amount of a product, you should calculate the percent yield (show your calculations).

• Complete the notebook with Conclusions, which includes a summary of the results and addresses any difficulties encountered, and then answer any questions which may have been assigned in the experiment.

An example of a notebook report with this general format follows.

Experiment 3: Preparation of Ethyl N-Methoxy-N-methylcarbamate
Date: 9-24-93

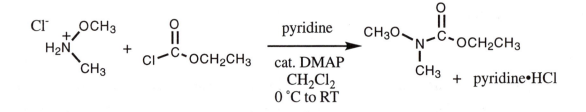

Purpose: Prepare carbamate derivative to use in conversion of organolithiums to ketones.
Reference: D. J. Hlast, J. J. Court Tetrahedron Lett. 1989, 30, 1773.

Table of Reagents

reagent	mol. wt./density	amount	mmol	equiv.	hazards
N,O-dimethylhydroxylamine hydrochloride	97.55 g/mol	25.56 g	262	1.0	hygroscopic
dichloromethane (dist. from CaH₂)		210 mL			toxic, volatile
4-dimethylaminopyridine (DMAP)	122 g/mol	1.04 g	8.5	0.03	highly toxic
ethyl chloroformate	108.52 g/mol 1.135 g/mL	25.0 mL	262	1.0	highly toxic
pyridine (dried over NaOH)	79.10 g/mol 0.978 g/mL	47 mL	576	2.2	toxic, stench

Procedure

The reaction was done in a 500 mL round bottom flask containing a large stir bar. The amine hydrochloride was weighed quickly in air and was then suspended in dichloromethane and cooled to 0 °C under inert atmosphere. Then dimethylaminopyridine and ethyl chloroformate were added; little or no reaction was evident at this point. Then pyridine was added by cannula over about a 5 minute period. The reaction mixture thickened considerably during the pyridine addition, to the point that magnetic stirring became difficult. After about 5 minutes at 0 °C, the mixture was allowed to warm to room temperature over about a 15 minute period. The suspension thinned somewhat and became more stirable, with some heat evolution noted also. The reaction mixture was allowed to stir until next lab period.

9-26-93

The suspension was cooled to 0 °C, then filtered through a coarse porosity fritted-glass funnel, rinsing the solid well with dichloromethane. The filtrate was then transferred to a separatory funnel and washed twice with 125 mL portions of ice-cold 3 M aqueous HCl. During the first wash, the aqueous phase became yellow, but the second was colorless. The organic phase was dried over magnesium sulfate, filtered and concentrated by rotary evaporation with the flask at or below room temperature. This yielded a clear, light brownish-yellow liquid. This was subjected to simple distillation under aspirator vacuum. This yielded 31.8 g of a clear, colorless liquid, with a boiling point of 74-76 °C at 43 torr. A sample was prepared for capillary GC by diluting a small amount in dichloromethane using the "paper clip" method. This showed the compound to be > 99% pure. GC-mass spectrometry was done, and ^1H and ^{13}C NMR were obtained at 360 and 90 MHz, respectively.

Percent yield calculations:

$$\frac{31.8 \ g \ product}{133 \ g/mol} \ \chi \ \frac{1000 \ mmol}{1 \ mol} = 239 \ mmol \ product$$

$$\frac{239 \ mmol \ product}{262 \ mmol \ limiting \ reagent} \ \chi \ 100\% = 91\% \ yield \ of \ product$$

Conclusions: This reaction was a good example of nucleophilic acyl substitution and vacuum distillation, and proceeded in excellent yield.

Capillary GC and GC-MS for ethyl *N*-methoxy-*N*-methylcarbamate (80 ˚C to 150 ˚C at 5 ˚C per minute).

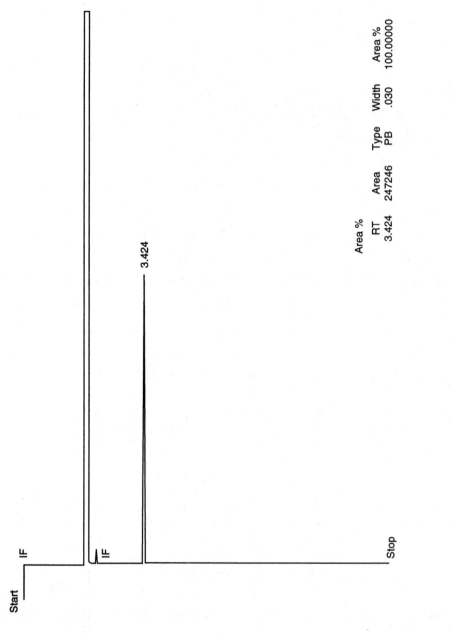

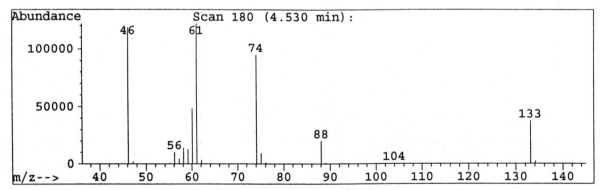

1H and 13C NMR spectra for ethyl *N*-methoxy-*N*-methylcarbamate (360 and 90 MHz, resp.)

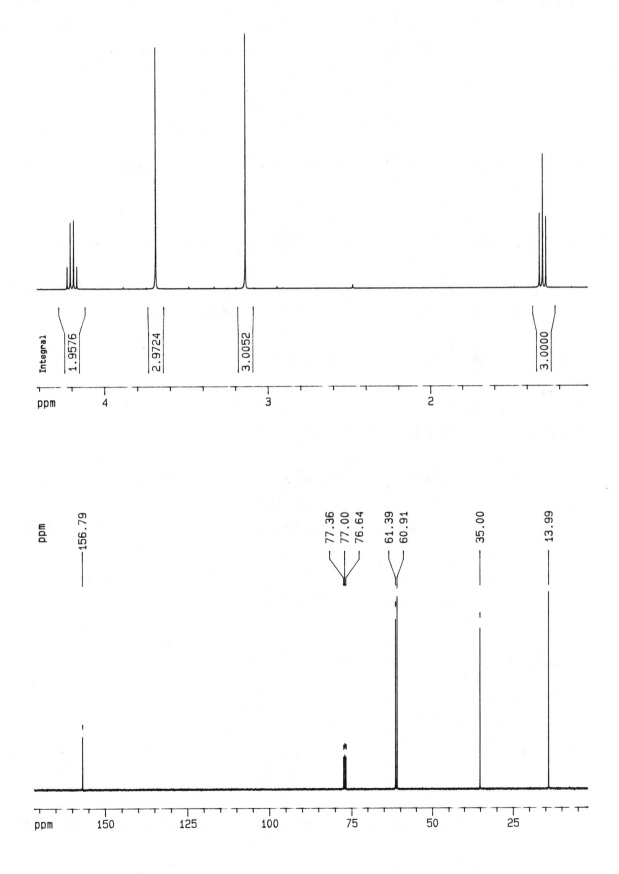

Characterizing products: You will be instructed as to which of the analyses/characterizations described below you should carry out in the course of a given experiment. However, it is instructive to consider why each analysis is being done and whether a different technique might be more informative. Keep in mind what concentrations are appropriate for each technique (see discussions of each technique).

Important Analyses/Characterizations:

Melting point (for solids) **or** observed boiling point (for liquids during distillation; include pressure!).

Thin layer chromatography (draw in the actual TLC plate, shade spots with pencil; specify solvent)

Proton NMR (with chemical shift scale printed and first-order multiplets expanded appropriately for determining coupling constants; see the discussion of proton NMR spectra).

Carbon NMR (with all relevant line frequencies noted in ppm).

IR (preferably neat, for liquids, or as KBr pellets, for solids; mineral oil mulls may also be used for solids, but the oil has three very strong peaks. Identify only meaningful peaks).

Capillary gas chromatography (always note the temperature program used, and identify peaks).

Capillary GC/mass spectroscopy (always note the temperature program used, and identify peaks).

PREPARATION AND ISOLATION OF PRODUCTS

Most syntheses require three distinct steps: (1) doing the reaction; (2) isolation of the crude product; and (3) purification of the crude product. Only occasionally will you encounter an especially simple procedure that manages to skip the second and/or third step.

Doing the Reaction

The first part of any given experimental procedure is designed to bring together the appropriate reactants in such a way that the desired reaction occurs in a controlled (not violent) manner. Generally, this involves dissolving the reactants in an inert solvent at a temperature where the reaction occurs at a reasonable rate. Reactions which would be too slow otherwise are heated, while reactions which would be too violent or unselective otherwise are cooled. Side reactions must be avoided or at least minimized. In cases where a reactant or product would be destroyed by oxygen or water in the air, this means doing the reaction under inert atmosphere. If moisture is deleterious, anhydrous (water-free) solvents must be used and glassware should be dried in an oven. If the time required for the reaction to be complete is not known, you should monitor the progress by TLC, GC, NMR, or IR. Which of these techniques is most appropriate will depend greatly on the particular reaction being done.

Heating and cooling reactions: Reactions are frequently done at other than room temperature. When heating is required, for reasons of control and safety this is usually done electrically rather than with a flame. There are two basic approaches to heating reaction flasks: (a) The flask can be immersed into a heat-transfer fluid in a dish on a stirring hot plate. The hot plate setting controls the amount of heat applied. Common fluids are water, mineral oil and flaked graphite. Water and mineral oil have the advantage of having well-defined and easily measured temperatures. Water is used only for heating at or below 100 °C, and because of the high humidity associated with a water bath, the reaction setup must be well-sealed when doing water-sensitive reactions. Mineral oil baths are useful up to 200-225 °C; above this temperature, the oil will produce excessive smoke. A special silicone oil may be used at higher temperatures. However, oil baths are messy and hot oil must be handled very carefully to avoid the possibility of burns. The liquid baths are used primarily when the flask temperature needs to be known with some accuracy. Interestingly, flaked graphite behaves as a fluid because the flakes simply slide out of the way when a flask is immersed. However, graphite is only a moderate conductor of heat; a significant thermal gradient exists from the bottom to the top of the bath, and the bath temperature is not well defined. Graphite is, however, a *much* better conductor than sand. Graphite can be messy when spilled, and can also make floors slippery. (b) Alternatively, the flask can be placed into a heating mantle, which is a hemispherical enclosure containing electrical heating elements. The amount of heat applied is controlled by a separate unit. Heating mantles are *never* plugged directly into an electrical outlet! Mantles are rugged and involve no liquids, but they make external measurement of the temperature difficult or impossible.

Knowing the reaction temperature: When a reaction requires careful temperature control, either a liquid heating bath is used or a thermometer is placed directly into the contents of the flask. The latter approach usually cannot be done with single-neck flasks, and requires the use of more exotic apparatus. When a somewhat less exact temperature control is required, it is very common to simply heat a reaction mixture to the boiling point. A water-cooled condenser is attached to the flask to prevent escape of the solvent vapor. The process of continuous boiling and condensation is called *reflux*. The temperature of a refluxing mixture will be slightly above the boiling point of the solvent(s) involved. By appropriate choice of solvent, the approximate reaction temperature is established.

Reactions under inert atmosphere: Many reagents are air sensitive, and reactions using such materials are best done under inert atmosphere. The most common inert atmosphere is nitrogen, though the more expensive argon is often used also. (The Earth's atmosphere is 78% nitrogen and 1% argon.) There are very few situations in which nitrogen cannot be used, probably the most common being when metallic lithium is used (it reacts with N_2 to form a nitride). The higher density of argon tends to keep it inside a reaction vessel should the cap need to be removed, for example, to add a solid.

Inert atmosphere is most conveniently established by means of a three-way stopcock (see Figure 2). This allows air in the reaction vessel to be removed by application of vacuum, then a 180° rotation of the stopcock allows immediate admission of inert gas. The system is said to be under passive inert atmosphere because, contrary to what many students assume, the gas is **not** being forced into the reaction vessel. Rather, excess gas simply exits through an oil bubbler, which prevents air from diffusing into the system. However, should the reaction vessel place a "demand" on the system (i.e., require entrance of gas to attain/remain at atmospheric pressure, e.g., when filling after vacuum or to compensate for gas contraction during cooling), the inert gas flow must be sufficient to prevent a backflow of oil and entrance of air. This is especially critical during filling of the container after application of vacuum, and is accomplished by increasing the gas flow temporarily and carefully (slowly) turning the stopcock. Once the reaction vessel is under inert atmosphere, a large flow of gas is **not** required, just enough to make up for demands on the system due to any cooling which occurs. This assumes there are no leaks in the tubing and connections. A simple leak test is to watch the bubbler when the gas is turned off completely; the oil level in the center tube should remain steady, not go back to the same level as the oil in the outer tube.

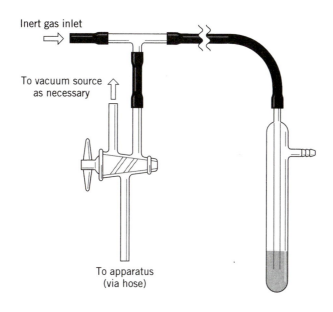

Figure 2: Three-way stopcock with inert gas and vacuum connections.

There are certain occasions when it is desirable to force inert gas into a vessel, such as for cannula transfers or when doing flash chromatography when compressed air is not available. A pinch clamp on the tube leading to the bubbler will allow this. Note that all connections must be capable of withstanding the pressure which is applied to the system, which should not exceed approximately 10 psi.

Note: Do not waste inert gas! You do not usually need a big flow of gas through your system, and no one wants to have to change gas cylinders. Experience shows that a gas cylinder can last for many months if waste is prevented. Turn off the gas when you don't need it, especially if it is used for flash chromatography columns.

About using Parafilm: It is used to seal edges that might otherwise allow gas transfer to occur (e.g., around the edge of a cap or over a septum). It is **not** used to cover openings on flasks or beakers. Never use Parafilm in areas where it will be exposed directly to nonpolar solvents (it dissolves) or to heat (it melts). The proper technique is to *stretch* the Parafilm as its applied, not simply press it on like clay.

Sources of vacuum: A good water aspirator will provide pressures as low as 20 ± 10 torr. In the absence of leaks or volatile solvents, the vacuum is limited only by the vapor pressure of water, which is roughly equal to the water temperature in °C. Aspirators are fairly rugged, typically lasting for a few years. It is an extremely simple device, but is susceptible to corrosion or clogging with particulates. It is important that your aspirator function well. Each one should be checked for maximum vacuum at the beginning of the course and monitored for loss of efficiency during the semester.

To obtain lower pressures than an aspirator will provide, you must use a mechanical vacuum pump. When working properly and in the absence of leaks, this will provide pressures of 0.1 torr or less. The pumping mechanism is somewhat similar to that in rotary (Wankel) engines, and the parts are bathed in a nonvolatile oil (for lubrication and, to some extent, for sealing the interior). However, when organic vapors are drawn into these pumps, they largely dissolve into the oil ("like dissolves like"), causing one or both of the following:

(a) The vapor pressure of the volatile solvent limits the vacuum attainable.

(b) The compound may promote corrosion (rusting) of the pump mechanisms (which are mostly cast iron).

The standard way to protect vacuum pumps from organic solvents is to install a *cold trap* between the pump and the source of organic vapors. This trap is cooled by immersion in either dry ice/isopropanol or (preferably) liquid nitrogen. The organic vapors are trapped as either liquids or solids and cannot reach the pump. Because of certain dangers involved in the use of liquid nitrogen (discussed below), *the vacuum system/trap assembly will be set up and taken down only by the instructor or with his permission.* However, for future reference you should be aware of the primary danger in working with liquid nitrogen:

Liquid nitrogen is cold enough that it can liquify atmospheric oxygen. If the trap is at liquid nitrogen temperature **and** not under vacuum, liquid oxygen will begin to condense. The danger is that liquid oxygen in contact with organics in the trap can detonate violently and unpredictably. For this reason, you should (a) never cool the trap before applying vacuum and (b) never shut off the pump without removing the dewar first.

Note: If you leave a liquid nitrogen-trapped vacuum system open, even with the pump running, liquid oxygen may condense! Always be certain either (a) to seal the system or else (b) to turn it off and remove the liquid nitrogen dewar before you walk away!

<u>Starting a pump</u>:

1. Start with a clean and dry trap.
2. With the system sealed (all valves closed, all hoses in place), start the vacuum pump.
3. Secure the dewar flask (which is empty at this point) around the trap; tighten the clamp firmly but neither too loosely or too tightly (dewars are expensive and will shatter if dropped).
4. Fill the dewar with liquid nitrogen (this will take a couple of minutes) or dry ice/isopropanol and wrap a towel around the top (this makes the coolant last much longer).

If the dewar is filled completely and you wrap the top with a towel, you can expect the nitrogen to last 10-12 hours, if only small amounts of organics are pumped into the trap. Dry ice/isopropanol should last somewhat longer. With your instructor's permission, you may be allowed to leave the pump on overnight if necessary, but you must fill the dewar at the end of the day, make sure it is well-insulated, and refill the dewar or turn off the pump first thing in the morning.

<u>Turning off a pump</u>:

1. First, lower the dewar and set it safely out of the way.
2. Vent the trap by removing a hose, and *immediately* shut off the pump.

<u>Protecting a pump from acidic gases</u>: Many reactions evolve acidic gases, particularly as HCl. Unfortunately, HCl is not condensed even at liquid nitrogen temperatures! It is best never to apply pump vacuum to a vessel containing HCl; aspirator vacuum can often be substituted in these situations. If it is necessary to expose your vacuum pump to an acidic gas, insert a drying tube filled with NaOH pellets before the trap.

Transferring liquids: The best technique for transferring a liquid to a reaction vessel depends on the properties of the liquid and the amount that is needed. For relatively large amounts (> 5-10 mL) of liquids which are not air or water sensitive, graduated cylinders are appropriate. However, hygroscopic or air-sensitive liquids must be transferred by either syringe or cannula techniques. The use of either syringes or cannulas requires that reagent bottles and reaction vessels be equipped with rubber septa.

<u>Using syringes</u>: The following is a step-by-step guide to using syringes.

(a) Choose a syringe whose capacity is appropriate for the job at hand, i.e., which will be at least one-third filled by the amount of liquid being transferred. This ensures that a reasonable degree of accuracy will be obtained. Be certain that the needle is securely attached. Note that luer-lock syringes require the needle hub to be twisted onto the syringe tip. This prevents the needle from coming loose during transfers. When using non-luer-lock syringes, be careful not to dislodge the needle accidentally. The following is the proper technique for using a syringe:

(b) If the liquid is air sensitive, remove air from the syringe and needle by drawing in inert atmosphere and expelling it to the atmosphere two or three times. Inert gas is obtained from a septum-capped vessel which is attached to an inert gas supply. An inert gas hose is easily connected to reaction flasks and other vessels by means of a "hose-to-needle adapter" which is connected to a short needle inserted through the septum (see Figures 3-5).

(c) To accurately measure the amount of liquid being transferred, it is important that no bubbles be present at the point of measurement. This is accomplished by holding the syringe vertical with the needle at top (see Figure 3) and drawing up more of the solution than is required. Any bubbles present are allowed to accumulate at the top of the syringe where they are ejected as the plunger is brought up to the mark.

(d) To prevent liquid from being thrown from the needle and/or air from reacting with material in the needle during transfer, the needle is raised above the level of the liquid in the reagent bottle and a small amount of inert atmosphere is drawn in to clear the needle of liquid.

(e) After removing the needle from the source vessel and inserting it through the septum on the destination vessel, the syringe is again held vertical while the inert atmosphere at the top is ejected, followed by the contents of the syringe. Note that this leaves the needle full of liquid during both the measurement and the expulsion. If gas remains in the syringe during either operation (i.e., if the syringe is not held vertical), a volume error equal to the needle volume may result.

If the remaining liquid is valuable and not compromised by its exposure to the reaction vessel, it should be returned to the original reagent bottle by turning the syringe vertical with the needle pointing down prior to expulsion. If the liquid is reactive or corrosive, the syringe and needle should be immediately rinsed with an inert solvent.

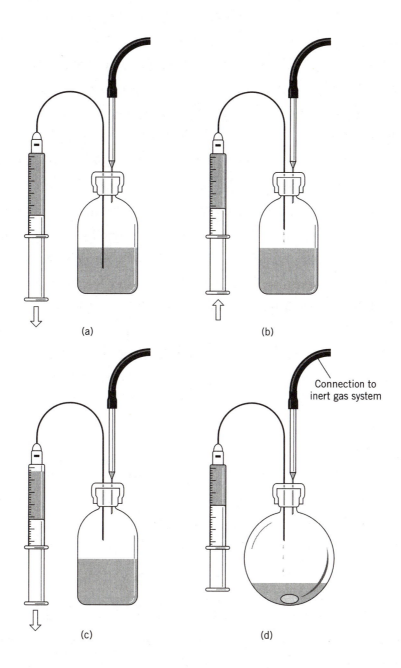

Connection to
inert gas system

Figure 3: Proper syringe technique. (a) Syringe filled with excess liquid. (b) Plunger pushed in to proper volume. (c) Small amount of inert gas drawn into syringe. (d) Liquid discharged with all gas expelled first.

<u>Transferring liquids by cannula techniques</u>: A cannula is simply a long tube by which liquid can be transferred between two septum-capped vessels. This is accomplished by inserting the cannula through the septum of the "source" vessel to below the liquid level, then pressurizing the vessel to

force the liquid through the cannula and into the "destination" vessel. Note that the pressure in the destination vessel must be less than that of the source vessel. In particular, any pressure in the destination vessel which develops as a result of the volume being transferred should be relieved. This is done through a second needle which is vented either to a bubbler or to the atmosphere (see Figure 4). Also, depending on the pressure used, the septum on the source vessel may need to be secured with wire or Parafilm to avoid having it pop off during pressurization.

Cannulas are generally made of thin-walled stainless steel tubing, and are typically 24-36 inches long. One or both ends have a sharp septum penetration point to simplify puncturing the rubber septa. There are several advantages to the cannula technique: (a) Exposure of the liquid reagent to the atmosphere is negligible, superior to that obtained with syringe techniques. (b) The volume transferred by cannulas is essentially unlimited, whereas syringes rarely exceed 50 mL in volume. (c) Cannulas are easier to clean than syringes; an appropriate solvent is simply forced through the cannula, or pulled through with aspirator vacuum. The disadvantages are that (a) an inert gas system capable of applying controlled pressure must be set up and (b) there is no intrinsic volume measurement with cannulas as there is with syringes. However, the use of a septum-capped graduated cylinder allows one to transfer measured amounts of liquids easily.

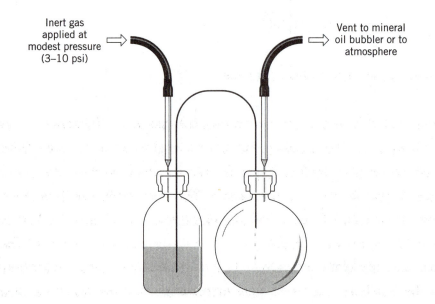

Figure 4: Liquid transfer by the cannula technique.

Cannula transfers can be done in conjunction with air-sensitive filtrations in one of two ways: If the liquid is easily decanted from the solid (e.g., large crystals present), the cannula can simply be directed into the free liquid while pressure is applied. In more difficult cases, the liquid can be passed through a Kramer filter, which is a tube containing a coarse porosity fritted-glass disk, sealed with a septum at one end and having a needle connection at the other (see Figure 5). With suitable care and speed, even hot filtrations can be accomplished with this technique.

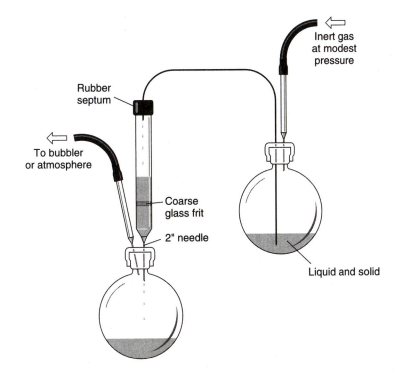

Figure 5: Cannula transfer with Kramer filtration.

Hint: In research labs, it is common to have certain solvents (e.g., THF) dried in a permanent still under an inert atmosphere. The easiest way to remove large amounts of solvent from such a still is as follows: Cap an oven-dried graduated cylinder with a septum and allow it to cool under vacuum (aspirator or pump) applied through a short needle. Then remove the needle without refilling with inert atmosphere. A cannula is inserted into the solvent reservoir of the still below the liquid level, and the other end is then inserted through the septum on the graduated cylinder. The vacuum will pull solvent over until the graduated cylinder is about 80% full, or until you lift the end above the liquid level in the reservoir. Be prepared to increase the inert gas flow to compensate for the volume removed from the still, especially at the end of the transfer when the residual volume in the graduated cylinder fills with inert gas.

About weighing: For small-scale reactions (< 1 g of starting material), weights should be determined to three or four decimal places. For larger-scale reactions, a top-loading balance which reads to two decimals may be used. However, choose a balance such that the uncertainty in the weight is always much less than one percent.

Note: Keep analytical balances absolutely clean. They cost at least $1,500 and should be treated with respect. *Do not transfer materials in the weighing chamber; always do the transfers outside the weighing chamber!* Use a top-loading balance for large amounts of materials (when reagent or product weighs over a few grams), and <u>always</u> when messy things are weighed out (e.g., silica gel).

<u>Weighing air-sensitive solids</u>: Some of the most problematic materials are solids which are air and/or water sensitive, and there are three different approaches, depending on how sensitive the compound is.

(a) If possible, use such materials as solutions of known concentration. This allows transfer by liquid techniques (syringes or cannula). Such solutions are often available from chemical companies (e.g., Aldrich).

(b) In cases where the compound is not rapidly degraded, the material can be handled quickly in air. In these cases, the best technique is to tare a dry container with an air tight closure, then quickly transfer approximately the correct amount of material, and reseal both containers prior to checking the weight. Material is then quickly added or removed to reach approximately the correct weight. It is often impractical, and unnecessary, to attempt to obtain exactly the desired amount. In most cases, the amounts of other reactants can be adjusted to achieve the desired ratios.

(c) If the material is rapidly degraded in air, the only alternative is to handle it in an inert atmosphere. This is done using a glove bag or a glove box. The former is a plastic bag with gloves built into it. The material to be transferred, as well as the <u>tared</u> container to which it will be transferred and other necessary items (spatulas, etc.), are placed inside and the bag is sealed. Residual air is removed, perhaps by aspirator, and the bag refilled with inert gas (nitrogen or argon). After the solid is transferred, both containers are tightly sealed before opening the bag. If necessary, it is possible to place a top-loading balance inside the bag to allow weighing under inert atmosphere. A glove box is a more elaborate enclosure which is permanently under inert atmosphere. Materials are brought into the box through an antechamber, which is essentially an air lock. The materials are placed inside, the chamber is sealed and air is removed by a vacuum pump. The vacuum is replaced with inert atmosphere, and the interior door is opened using gloves which are sealed into the box. After transferring the solid, remove the sealed containers through the air lock. Manipulations in either a glove bag or a glove box are difficult and clumsy, particularly in a glove bag. Glove boxes are more easily equipped with balances and other conveniences (even refrigerators), and they tend to have less residual air than glove bags. However, glove boxes are quite expensive ($10-40K) and require a lot of attention to maintain the inert atmosphere.

Note: Argon is significantly heavier than air, so if you tare a septum-capped container with air inside and subsequently use argon as the inert atmosphere, the container will weigh more afterward than the amount of solid accounts for. The error is approximately 0.5 mg per mL of volume.

Leaving equipment set up overnight: Usually, everything will be put away at the end of the lab period. However, in a few instances you will need to leave a flask or reaction setup in the hood overnight or longer. In these cases, you **must** label the apparatus with your name and the reaction being done. Condenser hoses must be securely attached before leaving any apparatus overnight; this means that you must wire the hose on to (1) the condenser, (2) the water spigot, and (3) the drain. Do not use water flows in condensers which are grossly more than you need - generally, just over a trickle is enough. No one appreciates floods!

Isolation of the Crude Product

Water workups: When the reaction is complete, the next step is generally to isolate the product in a crude (impure) form; this process is referred to as a "workup". If the desired products are fairly nonpolar, this often involves what is commonly known as a "water workup". Water is added to dissolve inorganic salts and/or very polar organics, and the nonpolar organic materials remain dissolved in the organic solvent (either present in the reaction or added afterward). Because most organic solvents are immiscible with (not soluble in) water, two phases form which are easily separated. Whether the organic phase is on top or bottom usually depends on the organic solvent used: halogenated solvents are more dense than water and form under a water layer, while all other organic solvents are less dense than water and will form on top. For volumes over about 5-10 mL, the separation is best done in a separatory funnel (see Figure 1). The lower layer is drained through the stopcock, but the upper layer is poured out the top afterward. It is a good idea to rinse the separatory funnel with a small amount of solvent to avoid transfer losses. For smaller volumes, careful pipetting serves to separate the phases reasonably well. This use of two phases to separate relatively nonpolar organics from very polar materials is referred to as "solvent extraction".

Note: (a) A water workup will generally not be used when preparing very water-soluble compounds, and (b) The most polar common organic solvents (methanol, ethanol, acetone) are infinitely miscible with water and cannot be used to perform solvent extractions.

If the desired product has appreciable water solubility, it will not necessarily dissolve efficiently into the organic phase. There are two approaches to improve the recovery of such materials from a water solution. By saturating an aqueous solution with a salt (usually sodium chloride), the solubility of organics will generally be decreased; this is referred to as "salting out" the organic material. Another approach is to increase the polarity of the organic solvent being used for the extraction. However, solvents which are infinitely miscible with water cannot be used. Polar organic solvents which have limited solubility in water are (most commonly) ethyl acetate or chloroform and (less commonly) 1-butanol. Chloroform is highly toxic, carcinogenic, and reactive under basic conditions, and its use should be avoided. Butanol is rather inconvenient to remove by evaporation (bp 118 °C), so ethyl acetate is the most convenient solvent for extraction of polar organics from water. Repeating an extraction with a new portion of organic solvent increases the efficiency of the recovery. Note that for *very* polar organics (e.g., sucrose = table sugar), extraction into any water-immiscible organic solvent is hopeless.

Emulsions: The most frequent problem encountered in solvent extraction is the formation of emulsions, which are mixtures of water and organic solvent stabilized by the presence of certain materials, especially powders or surfactant-like impurities. Vigorously shaking a separatory funnel makes emulsion formation more likely. For this reason, it is sometimes preferable to gently (rather than violently) mix the two phases especially in cases where emulsions are known to be a problem. Alternatively, simply being patient and waiting a while will often allow an emulsion to break up. In cases where a small amount of emulsion persists, it is best kept with the water layer. When doing a solvent extraction in a separatory funnel, you can usually remove most of the residual water in the organic layer by washing with a saturated sodium chloride solution last.

Drying: Once the desired product is in a volatile organic solvent, any water which may be present is removed by adding an inorganic drying agent. These are usually anhydrous metal salts which absorb water to form hydrates. The most common drying agent is anhydrous magnesium sulfate, but several others are also used. For reactions involving amines, basic drying agents like potassium carbonate are normally used. Cloudiness is often an indication that water is present in an organic solvent; however, clarity does not always mean a solvent is dry. Another indication of the presence of water is when a powdered drying agent "clumps" upon addition to an organic solvent. Based on these generalities, the drying agent is added until powdered drying agent remains "unclumped" and/or the solvent becomes crystal clear. In general, the more nonpolar the solution, the easier it is to dry by any method. The drying agent is removed by filtration prior to concentration by rotary evaporation. Remember to rinse the flask and drying agent with a small amount of solvent to avoid losing some of your product in the transfer.

Rotary evaporation: The solvent is usually removed by rotary evaporation (see Figure 6) and the equipment which does this is commonly referred to as a "rotovap". The basic idea is that application of aspirator vacuum (~ 10-30 torr) will greatly accelerate the evaporation process, but it also greatly increases the tendency of the material to "bump" or splash violently out of the flask. The chances of bumping are greatly reduced by rotating the flask while vacuum is being applied. This allows the liquid to wet a larger area of glass, which helps to give a very controlled evaporation. However, spinning too rapidly will cause the liquid to slosh, which makes bumping much more likely, while rotating too slowly does not wet the glass surface rapidly enough for smooth evaporation. Other factors which will increase the likelihood of bumping are applying vacuum too suddenly and heating the flask too quickly or too much. A very conservative, if slower, approach is to apply vacuum slowly (over 1-2 minutes) without heating until the outside of the flask gets cold, then immerse the flask in room-temperature water. The lowest boiling solvents

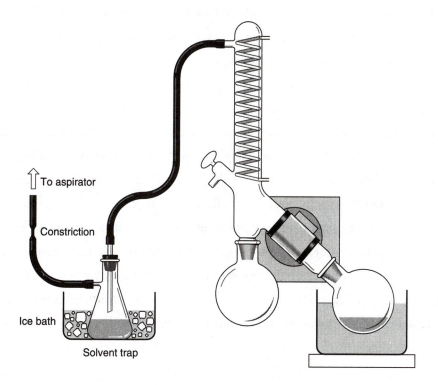

Figure 6: Rotary evaporation apparatus with separate solvent trap.

(ether, pentane, dichloromethane) are the most prone to bumping, as are mixtures in which solids are present. It is important to know your product's boiling point well enough (use the boiling point nomograph) that you are **certain** it will not be lost to evaporation on the rotary evaporator with the vacuum/temperatures you are using. Note that unless you are heating the flask above room temperature, none of the solvent will condense in a standard water-cooled rotary evaporator. An ice-cold trap connected to the vacuum source by a constriction (i.e., short section of capillary tubing) will give much more efficient condensation (see below).

Note: Leave the rotary evaporator clean for the next person. Try to avoid bumping your material in the first place by (a) not using flasks over half full and (b) applying vacuum slowly (over about 60 seconds or so). Also, traps are available to limit the contamination from bumping. To avoid possible contamination of your product, you should be sure the rotovap is clean before you use it. See your instructor for the best way to clean the rotovap.

Purification of the Crude Product

Purification is done using one or more of the common techniques (distillation, chromatography, or recrystallization), though on rare occasions an easier purification (e.g., simply washing with solvent) may be possible. Which technique is employed depends on the properties of the desired compound and impurities present.

Cleaning up

It is very important that each work area and the lab in general be kept clean. This includes balances, ovens, benches and hoods as well as your own glassware. Keep track of your glassware in the washing area so that it doesn't get put in someone else's drawer. Also, your glassware should be carefully cleaned, generally to the point of being spotless. You should remove any grease from the joints using dichloromethane or the ethanolic KOH bath (but see your instructor before using the latter). If you do use KOH/ethanol, do not leave your glassware in overly long (usually a few minutes to an hour will be sufficient). Alcoholic KOH will slowly etch and frost glass, so try to limit the exposure. **Always** wear safety glasses or goggles when working with KOH/ethanol because it is very caustic. The following guidelines apply to cleaning and handling your equipment.

• If you damage a piece of glassware, check with your instructor as to whether it may be repairable.

• Use grease on ground-glass taper joints ONLY in the following two situations:

(a) if the joint will be exposed to aqueous alkali (NaOH or KOH), or

(b) if you need to pull a very good vacuum (< 5 torr) on an apparatus. This will generally not apply when using aspirator vacuum because it can only reach perhaps 15 torr anyway.

• Do not put glassware in the oven if it is still wet with water, because someone else may need a very dry piece of equipment shortly thereafter and find the oven humidity at 100%! So rinse your things with acetone and let them air-dry a few minutes before putting them in the oven. Always check for star fractures before using round bottom flasks.

• Metal items (syringe needles) must be cleaned promptly to avoid corrosion. If you have syringed an even slightly corrosive material, rinse the syringe and needle <u>immediately</u> with an appropriate solvent.

• Always keep track of your magnetic stir bars and never let them go down a drain or be left in a waste bottle; they are very expensive and have a tendency to get away!

What to do if your glassware is not cleaned by the usual approach, soap and water followed by acetone? Here are four more drastic measures, listed in order of decreasing desirability:

(a) A thick slurry of Ajax (or Comet) will polish away almost anything. You have to be able to reach it with a brush, and be sure to rinse the Ajax away thoroughly when you are done.

(b) A solution of potassium hydroxide in isopropanol will remove many organics, especially silicone (high-vacuum) grease.

(c) Chromic acid (CrO_3 in H_2SO_4) will remove almost anything from glass, but this is a dangerous reagent, causing burns instantly on contact with skin. This method is recommended only in extreme circumstances.

(d) Traces of many organic compounds can be burned away in a glassblower's annealing oven, but if silicone grease is present, it will <u>permanently</u> cloud the glass surface.

Advice: Allow time to wash your glassware *before* you leave lab for the day; that way it will be ready to use the next day. Also, use appropriate judgment of how clean it needs to be. For example, an inorganic water spot (which will usually be on the outside anyway) will in no way influence your reaction! So it would be a waste of time to wash your flask and rinse it with acetone (which *will* usually mess up your reaction if you don't get it out).

PURIFICATION AND ANALYSIS TECHNIQUES

Some Practical Advice About Obtaining NMR Spectra

Cleaning and drying NMR tubes: Acetone takes a surprisingly long time (hours) to completely leave an NMR tube (even in the oven). For this reason, it is very common to see a huge acetone singlet in the proton NMR if you have recently cleaned the tube. So clean your tube at least a day before you will need it. If you must clean and dry one quickly, the best method is to briefly oven-dry it and then apply aspirator vacuum via a needle through a very small septum. Also, do not leave NMR tubes in the oven indefinitely because they can warp and become useless (a lot of the cost of these tubes is due to the straightness required).

Sample volumes: Most NMR spectrometers require 0.5-0.7 mL of liquid to function properly. This corresponds to a depth of 3.5-5 cm in a standard 5 mm OD NMR tube. Less volume than this may fail to fill the lock and/or observe coils in the probe, giving a poor lock signal (see below) and/or poor quality spectrum. The use of more solvent dilutes your compound unnecessarily and also wastes deuterated solvents, which tend to be expensive.

Sample concentrations: Modern FT-NMR spectrometers (i.e., those with a proton operating frequency of 200 MHz or greater) are far more efficient than the older CW instruments ($\leq$ 90 MHz). Samples as small as a few milligrams can give reasonable proton spectra without unduly long accumulation times, but you should try to limit TMS and water concentrations for such dilute samples to avoid the spectrum being severely dominated by these things. (For deuterochloroform, traces of water in the sample or solvent will appear at 1.5-1.6 ppm.) Very concentrated samples may give broader lines because of viscosity and also give a weaker lock signal. The "lock signal" is the deuterium NMR signal of the solvent (usually $CDCl_3$) which all modern instruments use to establish a field/frequency "lock", i.e., the instrument maintains a constant ratio of field to frequency and thus counteracts any tendency of the magnetic field strength to change during the NMR experiment. In addition, the lock signal is used to "shim" (adjust homogeneity of) the magnetic field so that all portions of the sample experience exactly the same magnetic field strength. The more homogeneous the magnetic field, the stronger the lock signal and the narrower the NMR peaks will be, down to a limit of somewhere around 0.5-1 Hz wide at half height. The strength of the lock signal will depend also on the deuterium concentration of the solvent (i.e., C_6D_6 or acetone-d_6 have a much stronger lock signal than does $CDCl_3$) and solvent concentration (i.e., stronger lock with more solvent/less sample). However, to obtain ^{13}C NMR spectra quickly with a good signal-to-noise ratio, fairly concentrated samples are desirable. With samples for which you intend to obtain both ^{1}H and ^{13}C NMR, solutions which are 20-30% sample/80-70%

solvent by volume strike a reasonable compromise, allowing you to obtain both proton and carbon spectra fairly quickly and yet providing a strong lock signal and fairly sharp lines.

Factors which contribute to loss of resolution in FT-NMR: You should be aware of the following considerations which can result in poor-quality NMR spectra.

Insufficient sample volume: To yield high-quality spectra, the solution must be at least somewhat more than enough to fill the receiver coil of the particular instrument you are using. This generally requires 3.5-5 cm (0.5-0.7 mL) of liquid in a 5 mm NMR tube, assuming the tube depth with respect to the spinner housing is correct. If in doubt, see your instructor.

Poor shims: *This is typically the limiting factor for most samples.* It is important to maximize the lock signal by adjusting some of the shim gradients. On modern superconducting magnets, you will usually adjust at least the Z and Z^2 gradients, and possibly also the Z^3 and Z^4 gradients. These controls adjust the current (and resulting fields) in small coils which supplement the main field, shaped and positioned such that it is theoretically possible to achieve a perfectly homogeneous field. Errors in the Z^2 and Z^4 gradients result in asymmetrical distortions, while errors in Z and Z^3 affect the width of the peaks. It is important to realize that the gradients are interactive: for example, adjustments of Z will change the optimum position of Z^2, and vice versa. Thus, if a significant change in one is made, the other should be re-optimized. The adjustments are made primarily based on the lock signal: improvements in field homogeneity will strengthen (narrow) the solvent's peak in the deuterium spectrum (which is what the lock really measures) and this will result in sharper peaks in the ^1H (and ^{13}C) NMR spectrum. If the instrument manufacturer describes how particular shim gradients affect the peak shape, this can be a very useful guide to shim adjustment, particularly in the final stages.

However, do not adjust any shims beyond those indicated by your instructor. Many of the gradients do not need adjustment between samples and will in fact seriously degrade instrument performance if improperly changed. In particular, do not adjust any shims which contain X or Y components (which affect the spinning side bands) unless you have your instructor's permission.

Note: Do not confuse shimming with tuning the instrument. Shims adjust the homogeneity of the magnetic field, while tuning refers to matching the transmitter/receiver coil in the probe to the appropriate pulse frequency. Poor shims affect the width and shape of peaks, while poor tuning will affect sensitivity drastically.

<u>Viscous solutions</u>: If samples are noticeably viscous, the lines will be broadened because the molecules will not be able to tumble rapidly enough to average their orientation with respect to the magnetic field. Such samples should be diluted and/or warmed to decrease the viscosity. Be aware that both field homogeneity and (especially) instrument tuning change significantly with temperature.

<u>Paramagnetic impurities</u>: Even trace amounts of materials with unpaired electrons can lead to broadening of the lines. This is most often observed with certain transition metal complexes. However, dissolved oxygen also causes this effect to a small extent (oxygen is really a diradical and therefore is paramagnetic), and deoxygenating the sample will narrow the line widths somewhat (but not usually enough to bother with). Any particle which is ferromagnetic (e.g., a speck of iron metal) will *severely* deteriorate resolution! It is usually a good idea to filter cloudy solutions before attempting NMR.

<u>Exponential multiplication</u>: This is a mathematical trick used to increase the signal-to-noise ratio (S/N) but which causes some loss of resolution. This is almost always required to achieve good sensitivity in carbon NMR spectra. In proton NMR, you should do this *only* if you would otherwise have poor S/N (which is not usually the case). Fortunately, for proton spectra we can reverse the effect to increase resolution at the expense of some S/N; this is referred to as <u>resolution enhancement</u>. To accomplish this, you multiply the FID by an *increasing* exponential. Sometimes this will give dramatic increases in resolution, but S/N will also be deteriorated. Exactly how this is done depends on the instrument; consult your instructor.

<u>Insufficient digital resolution</u>: The instrument generates your spectrum by simply connecting data points which are uniformly spread out over your spectrum, and "digital resolution" refers to the spacing (in Hz) between these data points. While a thorough discussion of Fourier transform NMR techniques is beyond the scope of this text, the main importance of digital resolution is that *you cannot accurately define a coupling constant which is less than twice the data point spacing.* The number of data points in the final spectrum will generally be one-half of the memory size used to store the FID, and they will be spread out over the entire spectrum (called the "sweep width"). Thus, the spacing (in Hz) between data points will be

$$\frac{2 \text{ x (sweep width in Hz)}}{\text{(memory size used to store FID)}} \quad \text{which equals} \quad \frac{1}{\text{(FID acquisition time in sec)}}$$

Note that if you	**the result will be**
Increase sweep width at fixed memory size	acquisition time decreases and spacing between data points increases
Increase memory size at fixed sweep width	acquisition time increases and spacing between data points decreases

However, the proper parameters are provided by the instrument and should not be changed without your instructor's permission.

Analyzing NMR Spectra

Analyzing proton NMR spectra: One of the major uses of NMR is to confirm that the product of a reaction has the expected structure. In this case, you must decide if the spectrum is consistent with the proposed structure. Examine the spectrum with the following questions in mind:

• Are expected <u>prominent</u> resonances (methyl singlets or doublets, downfield resonances) present?

• Are the chemical shifts, especially those downfield, consistent with the proposed structure?

• Is the expected coupling observed (at least for those resonances which do not overlap and which should show sufficiently simple coupling patterns)? Watch for unexpected long-range couplings ($\geq {}^4J$).

• Is the integration consistent? NMR only gives ratios unless you can assign a specific number of hydrogens to a particular resonance (e.g., 3H to a well-resolved methyl group, or 5H to a mono-substituted benzene ring). Be aware that vinyl and aromatic groups usually integrate somewhat smaller than expected relative to methyl or upfield alkyl protons.

In general, it is usually better to determine if the spectrum could possibly be consistent with the proposed structure rather than insisting that it appear as you might have expected.

Analyzing carbon NMR spectra: Because ${}^{13}C$ spectra should show a single peak for each type of carbon present, the number of peaks is the most important information present. In addition, the chemical shifts should be consistent. Any NMR text will have tables of typical chemical shifts for various structural features. However, it is often unnecessary to consult tables to evaluate a carbon spectrum. In general, ^{13}C peaks occur at chemical shifts about 20 times that of the attached protons (see: R. S. Macomber *J. Chem. Educ.* **1991**, *68*, 284-85). Because proton chemical shifts are more familiar and more predictable, this correspondence simplifies interpretation of the spectrum.

Reporting NMR Spectra

Reporting proton NMR spectra: When you wish to tabulate proton NMR data efficiently, as in a journal publication, proceed as follows:

• Go through the spectrum from right to left (upfield to downfield, or shielded to deshielded), noting the position and character of each discernible group of hydrogens in one of two ways:

A. Groups which **do** exhibit identifiable coupling patterns are reported as:

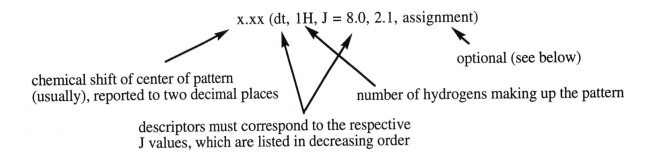

x.xx (dt, 1H, J = 8.0, 2.1, assignment)

chemical shift of center of pattern
(usually), reported to two decimal places

optional (see below)

number of hydrogens making up the pattern

descriptors must correspond to the respective
J values, which are listed in decreasing order

The following abbreviations are used to describe coupling patterns:

s = singlet   d = doublet   t = triplet   q = quartet   br = broad

(quintet, hextet, septet, etc., do not have abbreviations, so the whole word is used)

B. For groups which **do not** exhibit recognizable coupling, report as a **range** of chemical shifts covered:

x.xx-y.yy (m, 2H, assignment)

chemical shift
range of multiplet

optional

multiplet: contains complicated or unresolved coupling

Integrations are used to decide how many protons occur in each group encountered. Be aware that there is usually some error and that one proton may integrate to anywhere between 0.8 and 1.2 hydrogens, depending on the integration reference. For example, it is common for vinyl and aryl protons to integrate more weakly than saturated groups.

When reporting proton NMR data, assignments of the proton resonances are generally optional rather than required. However, your spectrum should be entirely consistent with the structure given. Also, proton-proton coupling constants should be reported to no more than one decimal place, and then only if your digital resolution justifies this. Otherwise, use no decimals on coupling constants.

Reporting carbon NMR spectra: When reporting carbon NMR data, list chemical shifts to no more than one decimal place, unless two lines are resolved only at the second decimal place. Depending on the quality of the spectrum, carbon chemical shifts are not usually reproducible to two decimal places.

Multinuclear NMR

When working with molecules which contain other NMR-active nuclei, the NMR spectra of these nuclei are often informative. This is especially true for "spin $1/2$ nuclei" because they tend to give narrow, well-resolved NMR signals. One advantage of multinuclear NMR is that deuterated solvents are not required: practically all solvents are transparent in multinuclear spectra! This allows reaction mixtures in common solvents (e.g., THF) to be observed directly, even though deuterated versions of these solvents are very expensive. Besides ^1H and ^{13}C, the most common and/or most easily observed spin $1/2$ nuclei are ^{19}F and ^{31}P. There are several possible differences between proton NMR and those of other nuclei:

Differences in sensitivity: There are two factors which contribute to sensitivity, natural abundance and the gyromagnetic ratio of the nucleus in question. The gyromagnetic ratio will be directly proportional to the observed frequency of a particular nucleus in any given magnetic field. For example, hydrogen (^1H) is 99.985% abundant and has the largest gyromagnetic ratio of any element, which makes it the most easily observed nucleus in the periodic table. On the other hand, ^{13}C is only 1.1% abundant and its gyromagnetic ratio is only one-fourth that of hydrogen. Because sensitivity is roughly proportional to the cube of the gyromagnetic ratio, these two factors make ^{13}C NMR approximately 6000 times less sensitive than that of hydrogen. Both ^{19}F and ^{31}P are 100% abundant and have fairly large gyromagnetic ratios, so these nuclei are among the most easily observed.

Differences in linewidths and couplings: The spin number of the nucleus often determines the quality of the spectrum. While spin $^1/_2$ nuclei give narrow NMR signals, nuclei with spins greater than $^1/_2$ generally give broader peaks. Examples of such nuclei are deuterium (^{2}H, $I = 1$), boron (^{11}B, $I = 3/2$), and ^{14}N ($I = 1$). Although many nuclei with spins greater than $^1/_2$ are observable by NMR, they generally give noisier spectra (broad lines are harder to detect than narrow ones) and more poorly resolved peaks. All spin $^1/_2$ nuclei follow the same coupling rules as proton (i.e., the "n+1 rule"), though such multinuclear couplings are often observed further than three bonds. Nuclei with spins greater than $^1/_2$ do not follow the proton coupling rules. For example, a single deuterium will split the directly attached carbon into three lines of equal intensity. The universal coupling rule which can be applied to all nuclei is

$$\text{Number of peaks } = \prod_i (2n_i I_i + 1)$$

where I_i = spin number of nucleus of type i coupled to nucleus being observed

               n_i = number of nuclei of type i which are coupled to nucleus being observed

Note that the spin number of the nucleus *being observed* does not affect the multiplicity observed in its own NMR spectrum, only the spins of nuclei *to which it is coupled*. Also, when more than one type of NMR-active nucleus is present, the total number of peaks observed will be the *product* of the individual splittings (in the absence of peak overlap).

Differences in chemical shift ranges: The range of chemical shifts are usually much larger than those observed for protons. This is also true of ^{13}C NMR (~ 200 ppm), which is about twenty times that of ^{1}H. To be certain that you have a large enough "window" to observe a multinuclear NMR signal, you should have some idea of what sweep width to use. Also, to correctly assign chemical shifts, you should be aware what compound is used for the zero reference point in the spectrum you are observing.

Three experiments in this text deal with aspects of multinuclear NMR: Experiment 4B (^{19}F), Experiment 7A (^{31}P), and Experiment 10C (^{31}P).

Distillation

Distillation separates compounds based on differences in boiling points. Variations of this technique are suitable for both large- and small-scale purifications, but the degree of separation depends on the difference in boiling point and the particular technique used. In all types of distillation, a liquid is boiled (vaporized) by application of heat and possibly vacuum, and the resulting vapor is cooled and condensed into a separate container. For this technique to be applicable to a given purification requires that the organic compound in question be stable to the heating required. However, the amount of heat required depends on the particular type of distillation chosen. Several varieties of distillation and the apparatus used in each are discussed below.

Simple Distillation: In this approach, the vapor is condensed a relatively short distance from the point of vaporization. Several types of apparatus are used to accomplish this; the particular one used will often depend on the scale (amount distilled). To efficiently separate compounds by simple distillation, very large differences in boiling points (≥ 50 °C) are necessary. Three types of apparatus for simple distillation are shown in Figure 7.

Kugelrohr distillation: Also known as bulb-to-bulb distillation, this simple distillation technique is especially useful for small-scale separation of a volatile compound from nonvolatile (or at least much less volatile) material. The compound to be distilled is placed in a round bottom flask and connected to a special bulb, which is usually connected in turn to a vacuum source (see Figure 7). In general, aspirator vacuum is used for compounds boiling between about 150 °C and 220 °C, while mechanical pump vacuum would be used for higher-boiling materials. The round bottom portion of the apparatus is placed in an oven and heated while being rotated to prevent bumping. The rotation can either be done manually or with simple mechanical oscillator systems which use air pressure or vacuum. The volatile compound distills into the bulb, which is usually cooled with dry ice to prevent loss of the material due to evaporation. The purified compound is recovered either with a bent pipette or (preferably) by turning the bulb upright and allowing the liquid to drain into a vial. The strength of this technique is the ability to distill small amounts of even high-boiling compounds with high recovery. Amounts as small as a few hundred milligrams can be purified with only moderate losses. The disadvantage of the technique is the low separation power: very large differences in boiling points (≥ 100 °C) are required to achieve good separation. One use of Kugelrohr distillation is to decolorize compounds purified by flash chromatography. Flash chromatography inevitably leaves a weak yellow coloration (most noticeable in polarimetry samples) which can be completely removed by bulb-to-bulb distillation.

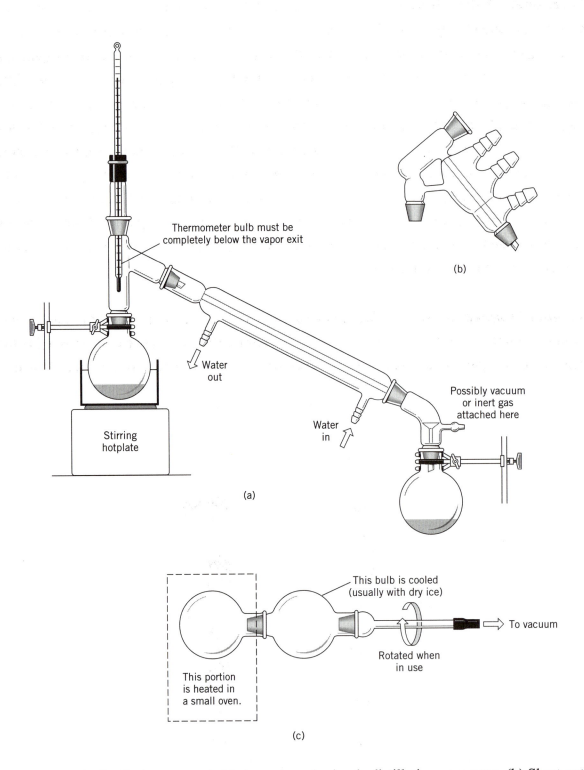

Thermometer bulb must be
completely below the vapor exit

(b)

Water
out

Possibly vacuum
or inert gas
attached here

Water
in

Stirring
hotplate

(a)

This bulb is cooled
(usually with dry ice)

To vacuum

Rotated when
in use

This portion
is heated in
a small oven.

(c)

Figure 7: Simple distillation apparati. (a) A macroscale simple distillation apparatus. (b) Short-path distillation apparatus. (c) Kugelrohr (bulb-to-bulb) distillation apparatus.

Fractional Distillation: In this technique, a column is inserted between the points of vaporization and condensation. The surface area of the column allows the vapors to condense and revaporize several times before being removed from the system. Each condensation/vaporization cycle allows the vapor to be further enriched in the lower-boiling component, with the higher-boiling component tending to run back down into the flask. Given large enough columns of the proper design, fractional distillation can give much better separations than simple distillation does. However, the process is slower and a certain amount of the mixture is required to wet the column surfaces and is lost. One of the most popular designs is the Vigreaux column (Figure 8). Various types of columns can exhibit widely different efficiencies; for example, a Vigreaux column 10 cm long adds approximately one additional theoretical plate of separating power. Higher efficiency separations can be accomplished using *spinning band* fractionating columns, microscale versions of which are commercially available.

Vacuum Distillation: The application of vacuum is often used to lower the boiling point of medium-to-low-volatility organics (bp > 200 °C) so that distillation can be accomplished at more convenient temperatures and/or without decomposition. The effect of pressure changes on

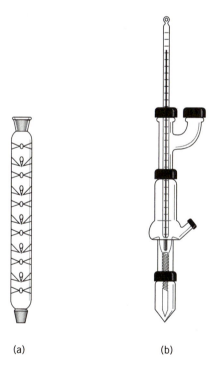

(a)   (b)

Figure 8: Vigreaux-type fractional distillation column (a) and microscale spinning band distillation apparatus (b).

observed boiling points can be estimated fairly accurately using a boiling point nomograph (Figure 10). In general, aspirator vacuum (20 torr) can be expected to lower a boiling point by 100-150 °C and pump vacuum (0.1 torr) by 200-230 °C. Vacuum distillation requires a fully sealed system and a reliable vacuum source. If one is to collect fractions during vacuum distillation, a special apparatus which can be rotated under vacuum is used (Figure 9). The complications of bumping and/or foaming are much more common in vacuum distillation. Because of this, it is especially important that boiling stones or vigorous stirring be used, and that the distillation flask not be over half full.

Figure 9: "Cow" assembly for fractional distillation under vacuum.

Steam Distillation: When dealing with organic compounds which are immiscible in water, steam distillation can be used to accomplish the distillation without heating above the boiling point of water. This takes advantage of the fact that for immiscible materials, the total vapor pressure of a mixture will be the *sum* of the separate vapor pressures. (This is in contrast to miscible mixtures, which obey Raoult's law and give a vapor pressure somewhere between the pressures of the pure compounds.) Water, of course, has a vapor pressure of 760 torr at 100 °C, while at this temperature a higher-boiling organic will have a much lower vapor pressure, perhaps only 10-30 torr. So when a mixture of water and an immiscible organic is heated, the mixture will always boil at least slightly below 100 °C and the distillate will contain both water and the organic compound (assuming the organic compound has any significant vapor pressure near 100 °C). The distillate will contain much more water than organic, according to the equation below, where the vapor pressures involved are those at the boiling point of the mixture. An older method of doing this was to pass steam into the mixture, but this was complicated and required an external steam source, which is less commonly available today. An easier method is to simply mix the organic compound with many volumes of water in a large flask and carry out the process using any of the simple distillation apparati. Once collected in the receiver, the organic compound forms a separate phase which is then removed and dried. This allows an organic material to be distilled at a temperature far below its boiling point, but there are restrictions: the organic must have a significant vapor pressure (say, $\geq$ 10 torr) at 100 °C and must be immiscible with and chemically stable to boiling water.

$$\frac{\text{moles of organic}}{\text{moles of water}} = \frac{\text{vapor pressure of organic}}{\text{vapor pressure of water}} \qquad \text{(Equation 1)}$$

Practical aspects of distillation: When attempting a distillation, you should particularly keep in mind the following:

Size of the flask: In preparing for a distillation, you will want to have the desired compound in a fairly concentrated form in a flask that is not overly small or large. If you use a flask which is too small, you run the risk that foaming or possibly bumping will physically carry over the material to the receiver without purification. Foaming and bumping are both much more likely when doing vacuum distillations. Overly large flasks will result in a loss of more material than necessary. Remember that the distillation requires wetting the interior of the apparatus, and the liquid required to do so will not distill, so is lost. A good compromise is to use a flask which has a volume between two and four times the volume of material you plan to distill. In most cases, dilute solutions are concentrated by rotary evaporation prior to distillation.

When to use vacuum: Before beginning a distillation, you should always know what boiling point to expect. Distillations are most conveniently done under conditions where the observed boiling point is between about 60 °C and 120 °C. Lower boiling points can result in loss of material due to poor condensation efficiency, and higher boiling points require too much heat to be applied, risking decomposition or even making the distillation impossible. Materials which boil above about 200 °C are almost always distilled under vacuum, and the expected boiling point at a given vacuum level is estimated from a boiling point nomograph (see Figure 10). The nomograph deals with three parameters: the compound's atmospheric boiling point (column 2), a particular reduced pressure (column 3), and the boiling point at reduced pressure (column 1). A straight line intersecting any two values allows prediction of the third. For distillations performed at less than about 10 torr, you should lightly grease all ground-glass joints by applying a small amount of high vacuum (silicone) grease to the upper one-third of the joint.

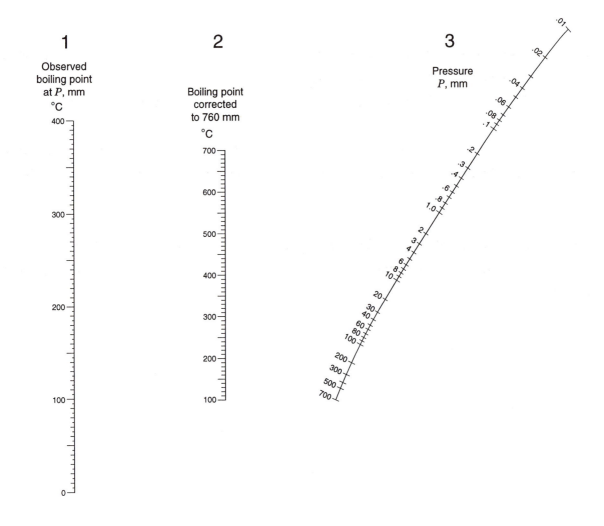

Figure 10: Boiling point nomograph for predicting boiling points at various pressures.

<u>Tips on distillations</u>:

(a) Vigorous magnetic stirring nearly always accomplishes the same thing as boiling stones, which are only rarely used in research. If you *do* decide to add a boiling stone, never add one to a hot solution!

(b) Be very careful with thermometers, as they are probably the most easily broken piece of apparatus. When doing a distillation, always install the thermometer last and remove it first when done. The correct position for a thermometer will place the entire mercury bulb slightly below the lowest path that the hot vapor can follow on its way to the condenser.

(c) Do not trust ground-glass joints to hold together unsupported! If you need to hold two joints together in a situation where they could come apart, a Keck clip (blue for 19/22 joints) will be very useful.

(d) Distillations always seem to take longer and require more heat than they should! It is important to insulate exposed glass surfaces on the "vaporization side" of the apparatus to speed up the process. Aluminum foil (shiny side in) or folded paper towels, secured with tape, work well.

(e) When doing vacuum distillations, always apply the vacuum first *then* heat as necessary to accomplish the distillation. **Never** apply vacuum to an already-hot liquid; it will foam or bump violently.

When a separation is too difficult and/or there is not enough compound for distillation, preparative GC is a good alternative if only small amounts of pure compound (< few hundred milligrams) are needed.

Gas Chromatography

Gas chromatography (GC) is a well-established technique for the analysis and/or small-scale purification of organic compounds. In all types of GC, the sample is rapidly vaporized in a high-temperature injection port (Figure 11) and swept onto and through a column with an inert carrier gas; the compounds are detected electronically after separation. Traditionally, GC columns were relatively short (4-12 ft) and wide ($1/8$-$3/8$" ID), packed with inert particles coated with a rather thick layer (2-20% by weight) of a nonvolatile organic liquid (called the *stationary phase*). These wide packed columns are necessary for preparative separations (i.e., where collection of multi-milligram amounts of compound is desired) and exhibit a separating power of about 100-1000 theoretical plates (TP); one TP is a unit of separating power approximately equal to that of a simple distillation. More recently, capillary columns have revolutionized the practice of analytical GC.

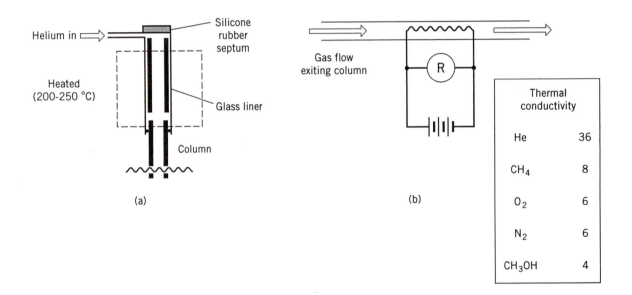

Figure 11: (a) Packed column injection port. (b) Schematic of thermal conductivity detector.

Preparative gas chromatography: Preparative GC is a powerful method for the purification of small amounts of organic compounds. This technique uses metal or glass columns packed with inert particles coated with the stationary phase. For preparative purposes, several aspects of the instrument are optimized as follows:

(a) The columns are shorter (3-12', or ~ 1-4 m) and wider ($1/4$-$1/2$" ID) than analytical columns.

(b) The particles are larger (80-100 mesh) than used in analytical packed columns, and are much more heavily coated (10-20% stationary phase by weight). As a result, at any given temperature, a preparative column will retain compounds more strongly (longer) than an analytical column would.

(c) For preparative work, the detector used must be nondestructive, and thermal conductivity detectors are used almost universally. These detectors monitor the heat conductance of the gas exiting the column. A crude schematic of a thermal conductivity detector is shown in Figure 11. The exit gas is passed over an electrically heated wire (called a filament), the temperature of which will depend on how quickly the gas can carry away excess heat. Because helium has the highest thermal conductivity of any gas (except hydrogen), pure helium keeps the filament cooler than will a mixture of helium plus an organic vapor. The resistance of the filament also depends on its temperature, so the thermal conductivity of the gas is easily monitored. While TCD detectors are much less sensitive than FID detectors (below), they are quite sensitive enough for preparative purposes, with much less than 1 mg of compound able to give full scale peak with the proper instrument settings. Another difference is that a TCD will detect air, water, carbon dioxide, etc., while FID detects only combustible organics.

As a compound of interest elutes (exits) from the detector, it is passed through a glass collector where it partially condenses back to a liquid (or solid, depending on its melting point). However, condensation efficiencies are only 40-60%, with the remainder of the material simply passing through as a gas or mist. Even extreme cooling in dry ice or liquid nitrogen does not significantly improve the collection efficiency. Collectors are designed to allow the liquid to pool for convenient withdrawal by pipette. If care is used in the process, materials from preparative GC are usually very pure.

Practical considerations:
Samples: Preparative GC uses concentrated solutions (50-80%) of the compound in a volatile solvent such as ether or hexanes. Because the solvent vapor may accumulate in the room, the more toxic solvents such as dichloromethane should be avoided unless very good ventilation is provided. Keep in mind that if compounds of longer retention time are present, either you will have to wait for them to clear the instrument or (if they have very long retention times) you may attempt two or more injections/collections of the desired compound before the higher-boiling materials elute.

Temperatures: The injection port and detector temperatures should be 200-250 °C; if the detector temperature is too low, the compounds can condense before eluting! The oven temperature will generally be between 100 °C and 200 °C, but not in excess of that recommended for the stationary phase in use. Exceeding the maximum temperature (T_{max}) of the stationary phase will result in baking off the more volatile portions of the phase plus any high-boiling compounds which may have been injected previously. For this reason, it is often a good idea to bake a column for 30-60 minutes just below its T_{max} prior to doing preparative GC.

Achieving Separations: In gas chromatography, separations are influenced primarily by the choice of the stationary phase and control of the oven temperature. A wide variety of stationary phases are available. However, in practice the common nonpolar stationary phases (e.g., SE-30 or similar) often give excellent separations. The oven temperature is adjusted to (a) achieve complete separation of the compound(s) of interest from all other materials (b) in as short a time as possible. Lower temperatures give better separations, but the compounds will require longer to exit the column. Temperature programming (i.e., increasing the oven temperature at a controlled rate during the chromatography, discussed below) helps greatly but is usually not available on preparative GC instruments. The final consideration is the amount of sample which may be injected, with small injections giving better separations than large injections. The amount which may be injected will depend on the size of the column (large columns can tolerate more compound) and the difficulty of the separation (easy separations can tolerate much more compound). Keep in mind that the peak shape will change depending on the amount injected: small injections should give symmetrical peaks, but large amounts of sample will give increasingly "shark fin"-shaped peaks (Figure 12). You should **not** attempt to control GC separations by changing the carrier gas flow. This is not effective and in the extreme could damage the instrument. The ideal carrier gas flow (50 mL per minute for $1/4$" columns, 200 mL per minute for $3/8$" columns) should be established when the instrument is installed and not changed thereafter.

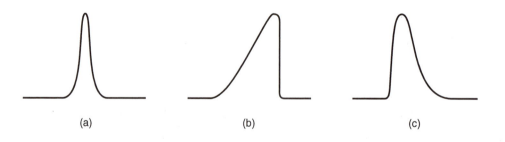

| (a) | (b) | (c) |

Figure 12: Common peak shapes. (a) Ideal peak shape. (b) Fronting due to column overload. (c) Tailing due to column aging and/or polar functional groups.

Capillary gas chromatography: A major breakthrough in analytical GC occurred with the development of capillary GC columns. These are extremely narrow (0.2-0.32 mm ID), long (10-30 m, ≈ 35-100 ft) tubes which are coated on the inside with a thin layer (0.25-1 μm) of an organic liquid. The column separates very small quantities (micrograms or nanograms), so capillary GC is used for strictly analytical purposes. The technique gives very narrow peaks (≈ 2-4 sec wide at half-height) and extremely good resolution (50,000-200,000 TP). While the molecules to be analyzed must be able to be vaporized without decomposition, capillary GC is much better suited for the analysis of larger, low-volatility molecules than are packed columns. There are three reasons for this: (a) capillary stationary phases are very thin and do not retain compounds as strongly; (b) capillary columns are better able to withstand the high temperatures required to get larger compounds through; and (c) capillary columns are made of fused silica, which is more inert than glass and helps to avoid decomposition of compounds. (See discussion of "Predicting GC retention times" below.) Probably the best phase for general work is the SE-54 column (also referred to by various manufacturers as HP-5 or DB-5), which has a maximum temperature of 350 °C, though there is rarely a need to use it above 300 °C.

There are several differences between capillary GC and the packed-column GCs used in Sophomore organic labs. Because capillary columns are so narrow, the gas flow through the column is less than 1 mL per minute. In order to get very narrow peaks, the sample must be applied to the column very quickly after injection. That is, in order to get peaks which are less than a few seconds wide, the compounds must get onto the column in <u>less</u> than this amount of time because some broadening will occur as the sample travels through the column. This is accomplished using a technique known as *split injection*, in which most of the sample is discarded for the sake of applying the remainder to the column rapidly. The ratio of total flow through the injection port to the flow through the column is called the *split ratio*, and is typically set at 40-100:1 (i.e., only 1-2% of sample is applied to the column). This is done with a cleverly designed injection port, the critical features of which are shown in Figure 13. Split injection greatly increases the resolution (separation) of the various sample components. While the technique is accompanied by some loss in sensitivity, this is not a problem because the detectors used with capillary GC (below) are very sensitive.

Capillary GC instruments are usually equipped with a *flame ionization detector* (FID). The gas exiting the GC column is passed through a tiny air/hydrogen flame, and the electrical resistance of the flame is monitored; a schematic of the detector is shown in Figure 12. The air/hydrogen flame has a very high resistance, and the helium carrier gas has no effect on this, so normally very little current flows through the circuit. However, when an organic compound elutes from the column and is burned in the flame, ions are formed which greatly increase the conductivity of the flame. The corresponding increase in current is recorded as a peak. FID detectors are very sensitive, able to detect as little as 10^{-10} g of most organic compounds. They also have a large linear dynamic range, which is the range over which the signal varies linearly with concentration of an organic analyte. FID detectors are typically linear over a 10^5 range of sample concentrations, which simplifies quantitation (measurement) of compounds in the sample. However, the FID cannot detect air, water, carbon dioxide, or other noncombustible materials.

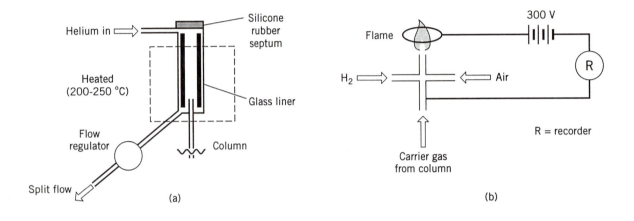

Figure 13: Schematics of (a) split injection port and (b) flame ionization detector.

Another difference is the use of *temperature programming*. When GC is done with the oven at a constant temperature, the lower-boiling compounds in the sample tend to elute quickly and be poorly separated, while the higher-boiling materials tend to take very long to come off the column and these peaks can be so broad as to be undetectable. An excellent solution to this problem is to begin the chromatography with the oven at a temperature low enough that the low-boiling compounds require several minutes to elute and so are well separated, then increase the oven temperature at a constant rate to accelerate the elution of higher-boiling materials. This has the effect of making the later peaks nearly as narrow as the early ones, all components tend to be well separated, and the analysis can be completed in much less time. In practice, the initial temperature is determined by experience (but see "Predicting GC retention times" below) and the rate of temperature increase (the *ramp rate*) is usually 5-10 °C per minute. For *very* difficult separations (e.g., diastereomers), a small ramp rate (1-2 °C per minute) is used and the initial temperature is chosen such that the compounds require 15-25 minutes to elute. Slower elutions result in rather broad peaks, while faster analyses give less separation. However, the resolution of capillary GC is so high that almost all analyses give complete separation of all components without our having to make any special efforts. Of course, separation between enantiomers is never observed unless perhaps the stationary phase contains chiral, nonracemic materials.

Setting up a capillary GC: If you happen to be setting up a capillary instrument for the first time, first set the desired linear velocity, then set the split ratio. The linear velocity is the speed at which the carrier gas travels through the column; this is controlled by the column head pressure and is determined by injecting methane (natural gas) at an oven temperature of 80 °C. For capillary columns (0.2-0.32 mm ID), a linear velocity of about 40 cm/sec is desired. For 30 m lengths, a 0.25 mm ID column requires a head pressure of about 15 psi, while a 0.32 mm ID column requires about 10 psi. The optimum split ratio can be determined by plotting the width of any non-solvent peak at various split flows; the lowest split flow (= lowest split ratio) for which a minimal peak width is obtained is best. This should occur at a split ratio of about 40-60:1, and yield peaks with half-height widths of 0.02-0.04 minutes. Because the flow through the column will be about 0.8-1 mL per minute, the total split flow will be around 40-60 mL per minute. Instructions on how to make these adjustments are found in your particular instrument's manual. However, do **not** make adjustments to the instrument without your instructor's approval.

Preparing capillary GC samples: Capillary GC is an extremely sensitive technique; typically 0.1% solutions (i.e., ≈ 99.9% solvent) are appropriate. More concentrated solutions (> 1%) often give poor chromatograms which show many impurities, even those present at only a few parts per million. A convenient way to prepare capillary GC samples from neat (solvent-free) organic liquids is to dip 5-10 mm of a clean, straightened paper clip into the liquid, then rinse this in a vial containing 1-2 mL of solvent (usually either dichloromethane or hexanes). More dilute solutions, such as reaction mixtures or chromatography fractions, are generally diluted 10-50:1 prior to analysis. Only very dilute solutions are suitable for direct analysis.

Note: Although the solvents used for capillary GC are not necessarily 100% pure (see below), they should be free of impurities with significantly higher boiling points. Otherwise, the chromatogram may exhibit extraneous peaks. In the relatively rare cases where this is a problem, simple distillation is generally sufficient to remove such impurities.

Note: When starting at sufficiently low temperatures, it is very common for solvents to exhibit more than one peak. This is because the solvent is present at such high concentrations that even minor impurities can be full-scale peaks. For example, dichloromethane will generally show four peaks if the initial oven temperature is around 40 °C; the major peak (due to CH_2Cl_2) is > 99% of the total. In my experience, all common solvents will be observed as more than one peak unless the initial temperature is high enough to merge them together. This is of little consequence because the recording integrator is generally set to ignore all peaks before about 2.5 minutes.

Predicting GC retention times: Unless you have previous experience with the compounds you are analyzing, it can be difficult to know what initial temperature/ramp rate/final temperature to use to get the compound to elute in a reasonable time. If the initial temperature is much too high, the compound can elute with the solvent and not be detected! Alternatively, if the initial and final temperatures are not high enough, the compounds being analyzed may never exit the column. If this happens, the high-boiling materials often produce extraneous peaks in subsequent chromatograms. Such peaks are easily recognized because (a) they will be significantly broader than other peaks in the chromatogram and (b) their retention times will not be reproducible if you repeat the analysis.

Because GC retention times are closely related to boiling points, which are related to molecular weight and polarity, the structure of the molecule allows you to roughly predict their GC behavior. With small molecules you must start with low oven temperatures or the compounds will elute so rapidly that they will not be sufficiently separated from the solvent peak. In general, unless the molecules contain functional groups, compounds with less than 7-8 carbons will elute too quickly to be separated from the solvent peak(s) at the lowest oven temperature conveniently attainable (40 °C). There are two solutions to this problem: (a) the sample can be prepared in a high-boiling solvent such as dodecane (bp 216 °C) or even hexadecane (bp 287 °C; mp 18 °C), in which case nearly all low-boiling materials can be analyzed (see "Exploring Further..." section of Experiment 5A); or (b) the sample can be analyzed on a packed-column GC, in which the thicker stationary phases used retain low-boiling materials to a greater extent. The upper limit to the size of molecule which can be analyzed by capillary GC depends mostly on the compound's stability to heat. In cases where stability is good, quite large molecules can be analyzed. For example, cholesterol ($C_{27}H_{46}O$, mol. wt. 386) elutes in approximately 10 minutes at an oven temperature of 300 °C.

Predicting retention times at various initial temperatures with a fixed ramp rate: The appropriate temperature programs and GC retention times of a wide variety of compounds can be estimated based on the behavior of the saturated hydrocarbons. Figure 14 shows the retention times of several hydrocarbons C_8-C_{14} using various initial temperatures, all using a ramp rate of 5 °C per minute on an HP-5 (SE-54) column, 0.25 mm x 30 m, with a 1 μm film thickness and 15 psi He pressure. This graph can be extended to functionalized molecules using the very empirical rules given below. This is done by assigning a "carbon equivalent" to other atoms such that the observed retention times fall on the same curve as the corresponding hydrocarbon. However, it turns out that the contribution of a heteroatom depends on the functional group it is in; for example, an ether oxygen is equivalent to less than one carbon while a carbonyl or alcohol oxygen is equivalent to more than one carbon (see Table 1).

To estimate a retention time of a compound using one of the initial temperatures below and a 5 °C per minute ramp rate, first determine the molecule's "carbon equivalent" as follows:

carbon equivalent = number of carbons + carbon equivalent of other atoms

Contributions from other atoms:

Each	**Atom**	of the following	**Type**		is equivalent to this	**Number of carbons**
	Oxygen		ether			0.33
			carbonyl			1.7
			hydroxy			2.4
	Nitrogen		amine	1°		2.2
				2°		1.2
				3°		0.81
	Chlorine					2.4
	Bromine					3.3

In addition to the atoms involved, add the following for these structural features:

aromatic ring	0.72
nonaromatic ring	0.5

Table 1: Carbon equivalents for other atoms in selected functional groups.

While the graph given in Figure 14 should allow estimation of appropriate temperature programs for most capillary GC instruments, it is not difficult to prepare a graph specific to your own instrument, in which case the accuracy may be sufficient to allow estimation of the size of molecules based only on GC retention times.

Note: The "Megabore" type of "capillary" columns are too wide (0.53 mm ID) to retain compounds as strongly as true capillary columns (0.2-0.32 mm ID). Instruments equipped with such wide-bore columns will give much shorter retention times than suggested by Figure 14, and also much less separating power (only a few thousand theoretical plates). Where possible, these columns should be avoided.

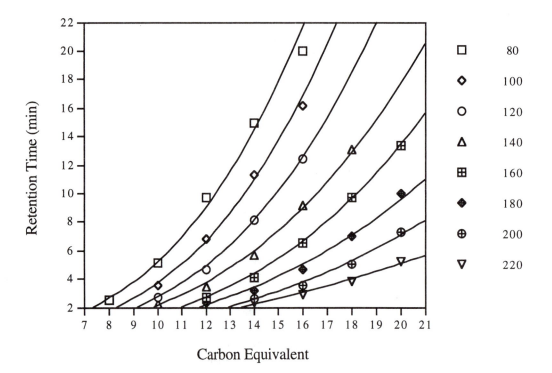

Figure 14: Retention times vs. carbon equivalents for SE-54 at various initial temperatures. In all cases, a ramp rate of 5 °C per minute was used.

Predicting retention times with various ramp rates at fixed initial temperature: Unfortunately, there does not appear to be a very accurate way to predict the effect of ramp rate on retention time. However, when the initial temperature is fixed, the retention time is a function of the ramp rate with *approximately* the following form:

$$\text{retention time} = \frac{\text{constant}}{\sqrt{\text{ramp rate}}} \qquad \text{(Equation 2)}$$

To *estimate* a retention time at any ramp rate, first use Figure 14 to estimate the retention time with a ramp rate of 5 °C per minute. Then solve Equation 2 for the constant and use this value to calculate (estimate) the expected retention time at any other ramp rate.

Quantitation of samples by gas chromatography: One approach to determining the amount of a compound present in a sample by GC is to compare the peak area obtained from injecting a given quantity of the sample (as a solution) with peak areas obtained by injecting the same volume of solutions containing known concentrations of the compound. This technique (the *external standard* method) is not usually the best because there is uncertainty in reproducing injection volumes (which are typically only 2 µL), peak areas are not necessarily a linear function of concentration, and it is necessary to know the volume of the sample solution accurately. A preferable method is to use an internal standard, which is an inert compound added to the mixture only for the sake of giving a reference peak. This is often done to determine yields from reactions, in which case the internal standard can be added either before or after the reaction is done. The procedure is as follows:

(a) An internal standard is chosen, usually a hydrocarbon like decane or dodecane. It must be inert to the reaction conditions and conveniently separable from the compound to be quantified and all other components of the reaction mixture.

(b) The detector's response to a given amount of the internal standard relative to the compound to be quantified is determined by preparing a sample with accurately known weights of the pure standard and the pure compound. These are diluted appropriately for GC (i.e., 0.1% for capillary GC) and analyzed. A *correction factor* is determined as follows:

$$\text{correction factor} = \frac{(\text{weight of compound}) \ (\text{area of standard})}{(\text{weight of standard}) \ (\text{area of compound})} \qquad \text{(Equation 3)}$$

(c) A known weight of standard is added to the reaction mixture (either before or after reaction), and the sample is analyzed by GC. The yield (in units of weight) is given by the following:

$$\text{weight yield} = \frac{(\text{weight of standard added}) \ (\text{correction factor}) \ (\text{area of compound})}{(\text{area of standard})}$$

$$\text{(Equation 4)}$$

Percent yields are obtained simply by dividing by the theoretical (weight) yield. Yields determined in this way are called "GC yields" as opposed to isolated yields of pure compound. GC yields are typically accurate to ± 5%, depending on the care taken to determine the values which went into the calculation.

Note: Before you may use any instrumentation (GC, GC-MS, NMR) independently, you must be checked out by your instructor to do so. When doing gas chromatography (including GC-MS), you must consider what and how much is present in your samples before injection, so that you do not load up the column or injection port with non-volatile or reactive materials. Learn not to inject more than you need (GC-MS requires higher concentrations than capillary GC does). Whenever you are doing manual injections, check the syringe tip for smoothness to ensure that you are not damaging the septum unnecessarily; see your instructor if in doubt.

Liquid Chromatography

Liquid chromatography techniques use liquid mobile phases and solid stationary phases. Because there is no requirement of volatility, liquid chromatography techniques are much more generally applicable to organic separations than is gas chromatography.

Thin Layer Chromatography: Thin layer chromatography is very useful for the rapid analysis of organic mixtures. Plastic, metal, or glass plates are purchased or prepared with a thin layer (0.25 mm) of finely powdered silica gel or (less often) alumina (called the *sorbent*). This material is held to the plate with a *binder*, which may be simply small amounts of plaster (hydrated calcium sulfate) incorporated into the sorbent. A fluorescent *indicator* is also usually present to help visualize UV-active compounds. The indicator absorbs UV light and emits visible light (usually green); the presence of a UV-active compound prevents the UV light from reaching the indicator in that spot. Thus, UV-active compounds appear as dull purple spots (simply a reflection of the visible part of the UV light source) on a green background. There are other methods available to visualize compounds which do not absorb UV light (discussed below).

TLC is used frequently in the following types of situations:

    (1) to monitor reactions for completion.
    (2) to check compounds for purity (GC or NMR also can be used for this).
    (3) to determine the appropriate solvent for a column chromatography.
    (4) to analyze the fractions collected from a column chromatography.

How to do thin layer chromatography:

Plates: Plastic-backed, UV-active TLC plates are convenient to use. These are hard enough that you can write on the surface with a pencil. These plates are commonly available as 20 x 20 cm sheets, so they must be cut with scissors to the appropriate sizes. Plates to be used for spotting column chromatography fractions are much wider (3-4 cm) than those used for other purposes (1.5-2 cm) such as reaction monitoring. It is convenient to divide the sheet into three equal sections, so the plates will be slightly over 6.5 cm high. Do not waste TLC plates; they are expensive!

Applying the compounds: It is important to apply the compounds as very small spots, which is done by very briefly touching a small-diameter capillary to the plate. Appropriate capillaries are prepared by drawing out a disposable pipette or a melting point capillary to the proper diameter in a flame. It works best to hold the tubing in the flame until it is somewhat softer than necessary, then remove it from the flame for about 1 second before pulling it into a capillary. Ideally, this will result in capillaries with an OD no larger than about 0.5 mm (about the diameter of paper-clip wire). Cut the capillaries into 5-10 cm lengths (you may want to make up a big supply at the beginning). Alternatively, you may use a short section of capillary GC tubing if some is available. In general, you must use fairly dilute solutions (1-5%) to avoid overloading the TLC plate; otherwise the spot(s) will be much larger and thus poorly resolved. The sample is best dissolved in a volatile solvent like dichloromethane or hexane, but any solvent with a boiling point less than about 80 °C will work well.

Choosing a TLC solvent: Unless you already know what solvent to use, first try a mixture of ethyl acetate in hexanes.* The goal is to get the compounds you are interested in into the correct range on the plate (R_f = 0.2-0.4 for the best separation; R_f is the ratio of the distance the compound travels to that the solvent travels). You will quickly learn what mixtures you will need for molecules of various polarities. For example, depending somewhat on structure, an alcohol will give an R_f of around 0.2 using 15% ethyl acetate in hexanes, while a diol (a dialcohol) might require 50% ethyl acetate in hexanes to get the same R_f. When dealing with multiple functional groups, the amount of ethyl acetate required to elute the compound is greater than the sum of the amounts needed for each of the individual functional groups. For very polar compounds like diols, one can also use methanol in dichloromethane. For compounds containing benzene rings, solvents containing benzene or toluene may give better separation. ***Caution:*** *Benzene is carcinogenic (causes leukemia) and must be handled in a hood while wearing gloves. Toluene is often substituted for benzene for this reason.*

* "Hexanes" refers to the commercially available mixture of six-carbon alkanes, and consists mainly of hexane (~ 70%) and methylcyclopentane (~ 30%).

Visualizing the spots: In many cases, not all of the spots on the plate will show up with a single visualization technique. Fortunately, you can apply three visualization methods to the same TLC plate and thereby get more information than any one technique would give. However, you must do it in the order given. First, UV light is used to show up aromatics and other UV-active compounds. Then the plate is exposed to I_2 vapor by placing it in a jar containing iodine and some silica gel. Finally, compounds are visualized by dipping the plate or (preferably) spraying it with a staining reagent followed (usually) by heating. If the dipping technique is used, the plate is immersed in the reagent *very* briefly (< 1 sec) and immediately dried on a paper towel prior to heating. Depending on the compounds being analyzed, any one of several stains may be used. The composition and behavior of some of the most common stains are given in Table 2 below. All require heating at approximately 100-150 ˚C, except permanganate, which is not usually heated.

Stain	Composition	Results
phosphomolybdic acid (PMA)	5% PMA (w/v) in ethanol	good general stain, compounds appear as blue spots
vanillin	6% vanillin (w/v), 1% H_2SO_4 (v/v) in ethanol	good general stain, compounds appear in various colors
permanganate	1% $KMnO_4$, 6% K_2CO_3, 0.1% NaOH in water	alkenes and easily oxidized compounds appear as yellow spots (no heating usually required)

Table 2: Common visualization reagents for thin layer chromatography.

Iodine tends to show up unsaturated materials fairly strongly, while PMA is excellent for alcohols and some other compounds. There are many other stains known to work with various compounds; vanillin/H_2SO_4/heating shows up many compounds, potassium permanganate is useful for visualizing alkenes, and tungstophosphoric acid (TPA) is sometimes useful for certain compounds.

Silver nitrate/silica chromatography: Certain materials differing in unsaturation or substitution around a carbon-carbon double bond can be separated on silver nitrate-treated silica, sometimes even when no separation is apparent on regular silica TLC. For example, α-pinene is well separated from β-pinene on $AgNO_3$-treated silica, but not on silica gel alone. The double bond will be bound to silver part of the time, and compounds with more "exposed" (less substituted) double bonds bind more strongly and thus move more slowly on the plate. In the case of the pinenes, the β isomer has the lower R_f on silver nitrate silica gel. However, the presence of silver makes visualization much more difficult (UV and I_2 are useless; permanganate and PMA function, but more weakly than normal). Divalent sulfur compounds (sulfides and thiols) bind very strongly to silver (I) and are easily separated from other materials.

(+)-α-pinene

(-)-β-pinene

Column Chromatography: Column chromatography is probably the most general (i.e., universal) method for the purification of organic compounds. The technique involves separation of individual compounds from a mixture by passing it through a column (Figure 15) packed with silica gel (or, less commonly, alumina). After the mixture is applied to the column, an appropriate solvent is passed through the column until the compound(s) of interest have *eluted* (exited at the bottom). Because different compounds will usually adhere to the silica to different extents, they travel at different speeds down the column. For example, compounds which spend more time attached to the silica travel more slowly than compounds which have little attraction for the silica. How quickly a given compound moves can be controlled by changing the solvent polarity: solvents

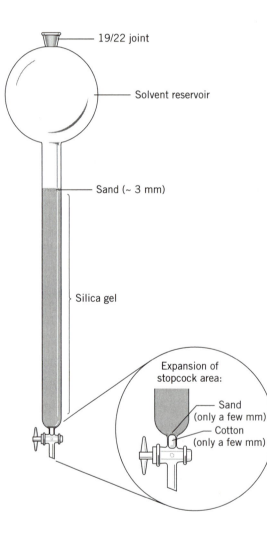

Figure 15: Diagram of column chromatography apparatus.

which are more polar cause compounds to travel faster. This is because compounds are attracted to silica by polar-polar interactions between Si-OH and Si-O-Si groups in the silica and polar functional groups in the organic compounds. Hydrocarbons, which have no polar functional groups, generally have little or no attraction to silica and travel very quickly. Other functional groups have more affinity for silica and compounds containing those groups travel more slowly to one degree or another. The solvent for a given separation is chosen based on the results of silica gel TLC: the proper solvent is that in which the compound has an R_f of approximately 0.25-0.3. However, if a small column (< 15 g of silica) is being used, a solvent which gives a TLC R_f of 0.15-0.2 is more appropriate. As the chromatography proceeds, the liquid exiting the column is collected in test tubes (called *fractions*) and analyzed (usually by TLC) to determine which fractions contain the desired compound(s). All of the fractions containing a desired compound in pure form are combined, and the solvent removed by rotary evaporation.

The most useful type of column chromatography is known generally as "flash" chromatography because it is faster than gravity-eluted columns. The basic idea is that by using pressure (5-10 psi) to force the solvent through the column, silica gel of smaller average particle size (260-400 mesh) can be used and still maintain adequate flow rates. When gravity alone is used, a larger average particle size (~ 70-230 mesh) must be used or else the column will elute too slowly. Because smaller particle sizes give better separations, flash chromatography provides rapid separations with better *resolution* (separating power). The technique was worked out mostly at Columbia University in 1978 (see W. C. Still, et al. *J. Org. Chem.* **1978**, *43*, 2923). However, many workers prefer longer, narrower columns (length:width ≈ 15:1) than those described by Still.

Practical guidelines for doing flash chromatography:
Amount of silica gel to use. Usually an amount of silica is used which is 50 times (by weight) that of the compound you will apply to the column. For very easy separations 10-30:1 is sufficient, while difficult separations may require 100:1. For example, if your reaction's theoretical yield was 300 mg, you could use 15 g of silica for a separation of average difficulty, and maybe 3 g for a very easy separation or 30 g for a difficult separation. The anticipated difficulty of a separation is based on the difference in TLC R_f between the desired compound and the nearest impurity (called ΔR_f): easy separations have $\Delta R_f > 0.4$, while difficult separations have $\Delta R_f < 0.1$.

What solvent to use. Use a solvent which gives your compound an R_f value of 0.2 to 0.3 on silica gel TLC; usually a mixture of ethyl acetate in hexanes will work. However, if you find another solvent which gives a better separation of your product from impurities <u>and</u> the correct R_f range

(0.2-0.3), use that solvent. For small columns (< 15 g of silica) it is best to use a somewhat less polar solvent than indicated by TLC; using only about two-thirds as much of the polar solvent as suggested by TLC is recommended.

Total amount of solvent required. You will need about 7-10 times more solvent (in mL) than the amount of silica gel used (in grams), assuming that your compound has an R_f of about 0.2-0.3 in that solvent and that you do not waste very much. This does not include the solvent required to pack the column.

Size of forerun. The volume of solvent which can be safely eluted before beginning to collect fractions is referred to as the "forerun". On columns using less than 15 g of silica, collect no forerun. On columns with 15 g or more, you can safely collect at least a forerun volume (in mL) equal to the weight of silica you used (in grams), and sometimes a few times this amount (especially on large columns). But err on the side of caution.

Size of fractions. Generally, fractions should be one-third to one-sixth as large (in mL) as the amount of silica gel (in grams). Smaller fractions can be beneficial for materials which are poorly separated, but you must keep the total number of fractions to a manageable number.

How to run a flash column step-by-step:
1. Choose a column which will conveniently hold the amount of silica you will need, with additional room in either the column or the reservoir (if present) for solvent. Place a small (~ 3 mm dia.) plug of cotton at the base of the column (all the way into the narrow portion just above the stopcock), then cover the cotton with a *little* sand (~ 3-5 mm).

2. In a hood, measure out the silica you will need *by volume* into a tared beaker or Erlenmeyer flask; the density is about 0.5 g/cc for well-settled flash chromatography silica. You may check the weight using a top-loading balance only (accuracy is not extremely important here). Do not breath silica dust, and clean up any that you spill. There are **two** common methods for packing the column:

   To **slurry pack** the column: Add some of the solvent you will be using (based on TLC R_fs) to the silica until a stirable suspension results (add just enough so that the mixture is thin enough that the air can escape), then pour this suspension into the column (a funnel helps here). Rinse the beaker once with a little solvent to get most of the silica transferred. Cautiously apply air pressure to pack the silica and lower the solvent to only a little above the silica. Do not allow air to enter the silica at any time. Rinse down any silica on the column walls with a little solvent, then sprinkle a layer of sand (~ 2 mm) on top of the column.

To **dry pack** the column: Using a powder funnel, pour the silica into the column, followed by several volumes of solvent. Cautiously apply air pressure to force the solvent through the silica. You should be able to observe two "solvent fronts" traveling down the column; the faster one will be a mixture of solvent and air, while the slower one will be more translucent and indicate the air-free zone. Continue the elution, adding additional solvent if necessary, until the slower solvent front reaches the bottom, indicating that the air within the silica bed has been expelled. Lower the solvent to only a little above the silica, but do not allow air to enter the silica at any time. Sprinkle a layer of sand (~ 2 mm) on top of the column.

Note: The silica must be within the *narrow* portion of the column and **not** in the reservoir portion! Otherwise you must scoop the excess out with a long spatula and rinse down the walls again.

3. Lower the solvent level until it just matches the silica height, then apply your compound using a disposable pipette. You should apply liquids neat (i.e., no solvent) if possible; otherwise use the least amount of the least polar solvent which will dissolve the material. (For solids which are not easily dissolved in a nonpolar solvent, see your instructor.) Then lower the liquid level again until it matches the top of the silica. Rinse the container which held your product once or twice with *very small* portions of solvent, applying each portion to the column and lowering the level each time. Then rinse the compound into the silica with several small portions of pure solvent.

4. Once you are **certain** that the compound is well within the silica, carefully add solvent by pipette until you can safely pour solvent into the reservoir without disturbing the silica. Place a graduated container under the outlet to measure the forerun (if any). Then clamp the gas inlet adapter onto the column and slowly apply air pressure to speed up the flow. (The flow should be substantial, with no more than 30 seconds per fraction required.) You can stop the flow using the Teflon stopcock at the bottom of the column when changing from one fraction tube to another. Make sure you keep track of the order in which you collect fractions! Do not let the column go dry before the desired compound has eluted (stop the chromatography long enough to do TLC if in doubt).

5. Analyze your fractions, usually by TLC. Keep in mind that, depending on your ability to visualize the compounds, the fractions might be a little dilute for TLC. So when spotting fractions, it is best to apply a given fraction **many times** (10-20) to the spot corresponding to that fraction (labeled on the plate with a pencil). Finally, when you have identified which fractions contain your

compound in sufficient purity, combine those fractions in a tared round bottom flask and remove the solvent using the rotovap. For volatile materials (bp < 200 °C), do not rotovap longer than necessary, and do not heat the flask above room temperature.

Hint: Upon standing 10-20 minutes in the hood, fractions containing substantial amounts of compound can usually be identified by a ring of liquid or solid 5-10 mm above the surface of the liquid.

Recrystallization

Recrystallization is a powerful method for the purification of solids, and is suitable for both very large scales (kg) and fairly small scales (≥ 30 mg). The technique basically involves a solid being dissolved in an appropriate solvent and caused to crystallize by one of several techniques. Because crystallization is a rather shape-selective process, the growing crystals tend to exclude impurities. The crystallization should occur slowly rather than rapidly. Extremely rapid crystallization is referred to as precipitation and usually results in little or no purification. The most common crystallization techniques are described below.

Hot/cold recrystallization: This is easily the most common type of crystallization and relies on the fact that hot solvents nearly always dissolve more solid than cold solvents do. The solvent is chosen such that the compound of interest will not be highly soluble in it. Ideally, the solid to be recrystallized would be reasonably soluble (say, ≥ 40 mg/mL) in the hot solvent, but rather insoluble (say, ≤ 10 mg/mL) in the cold solvent. "Hot" means within about 5-10 °C of the solvent's boiling point, and "cold" refers to either room temperature or 0 °C. Often the most difficult part of such a crystallization is finding the appropriate solvent. Because many organic solids are very soluble in medium polarity solvents (e.g., ether, dichloromethane), in many cases either extremely polar or extremely nonpolar solvents are required to meet the solubility requirements. Ethanol and methanol are polar enough that many organics will not be highly soluble, and small amounts of water can be added to decrease solubility even further if necessary. On the other hand, alkanes are so nonpolar that many organics will have only limited solubility. In either case, one would hope that heating would result in a significant increase in solubility, and that upon slow cooling crystals would form. The process involves the following steps:

(a) The compound is dissolved in somewhat more than the minimum amount of a hot solvent. This is usually done by **slowly** adding solvent to the crystals in a hot flask with good stirring. If the addition is too rapid or if the flask is not hot enough, it is very easy to add too much solvent.

Due to residual solubility, this often results in little or no compound recovered after cooling. If this happens, you must evaporate some of the solvent and repeat the heating/cooling cycle.

(b) If there are impurities which are not soluble in the hot solvent, the solution is filtered *while hot*. In practice this is often difficult because a significant amount of cooling is usually inevitable during the filtration. The danger is that the compound could crystallize rapidly in the filter and be rather bothersome to recover; in addition, recrystallization is still necessary. Fortunately, insoluble impurities are not often present. However, it is sometimes hard to tell during the heating phase if any remaining solid is the desired compound or an insoluble impurity. When in doubt, try to determine if additional solvent appears to be dissolving the solid or not. **Note:** Activated carbon is sometimes added to remove colored impurities, in which case a hot filtration will be required.

(c) The solvent is allowed to cool slowly to room temperature, possibly followed by cooling in ice-water. During this time, crystals should form slowly. Sometimes seeding (adding a tiny crystal) or scratching the inside of the flask with a glass rod will be important to initiate crystallization. Allow the flask to stand a while at the lower temperature; the crystallization often requires several minutes or more to be complete even after the temperature is no longer dropping.

(d) The crystals are filtered off, washed with a small amount of fresh solvent (sometimes ice-cold) and dried (either in air or under vacuum). Controlling the amount and temperature of the solvent limits the amount of solid which will be lost to solubility. For very large crystals, decantation (rather than filtration) may be used.

(e) Sometimes the filtrate (liquid which went through the filter) is concentrated by evaporation until a "second crop" of crystals is obtained. These may be less pure than the initial batch.

"Layering" recrystallizations: Another technique is to dissolve the solid in somewhat more than a minimum amount of a solvent in which the compound has a fairly high solubility, and then carefully apply a layer of a less dense solvent in which the compound has a low solubility. This is usually done in a moderately narrow container such as a vial. With care, two distinct layers will result. If the vial is left undisturbed, mixing occurs slowly over the course of several days, resulting in crystallization of the compound. It may be helpful to store the vial in a freezer during the mixing period. The success of the technique depends on:

(a) Correctly judging the amounts of each solvent and the compound's solubility in each solvent such that crystallization will occur upon mixing. Of course, the solvents must be miscible, and the top solvent should be less dense than the bottom one.

(b) Care during the layering process. Rapid mixing usually results in precipitation with no purification.

Dichloromethane is often a good choice for the lower layer because it usually dissolves large amounts of solids and is rather dense (*d* 1.43 g/mL). Common choices for the top layer are hexanes or possibly ether, since solids are often much less soluble in these solvents, and they have low densities (0.67 and 0.71 g/mL, respectively).

Diffusion recrystallization: In this technique, the compound is dissolved in somewhat more than a minimum amount of a solvent in which it is quite soluble. The vial is placed inside a larger container containing a volatile solvent in which the compound is much less soluble, and the larger vessel is sealed. If the solvents are properly chosen and if the container is left undisturbed, the volume in the center vial will increase slowly and crystals will form over a period of 1-2 days. One combination which works well is dichloromethane/ether in the inner/outer vessels, respectively. However, the choice of solvents is critical because not all possible combinations work. For example, the dichloromethane/pentane combination results in little or no diffusion.

EXPERIMENTAL PROCEDURES

Experiment 1: The Stereochemistry of Commercial 2,6-Dimethylcyclohexanone: An Application of Capillary Gas Chromatography

Capillary gas chromatography is an extremely high-resolution, fast and sensitive analysis technique, but has not been routinely employed in undergraduate organic laboratories. The following is one example of how capillary GC may be used to solve a stereochemistry problem which would be quite difficult by other common techniques. The ketone 2,6-dimethylcyclohexanone is available commercially as a "mixture of isomers" with neither the cis:trans ratio nor the identity of the major isomer stated. Although these isomeric ketones are easily separated by capillary GC, this alone does not allow one to assign the stereochemistry. However, upon reduction with sodium borohydride, each ketone produces a different number of alcohol products. Since all the alcohol isomers are easily separated and quantified, you can use this information to identify the original ketone stereochemistry. It is left to you to determine how many alcohol products each ketone is expected to give, but you should expect that the reduction reaction will occur to some extent at both faces of the ring (see below). *Note:* Enantiomers are **not** separated on nonchiral GC columns, but diastereomers almost always are.

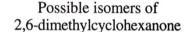

Possible isomers of
2,6-dimethylcyclohexanone

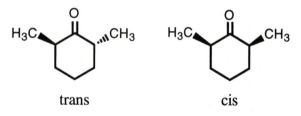

trans   cis

Expect H⁻ attack to occur to some extent at both faces

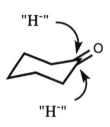

Procedure

If not provided by your instructor, determine the ketone isomer ratio of Aldrich 2,6-dimethylcyclohexanone by capillary GC analysis (Note 1), with the program set from 80 °C to approximately 96 °C at 2 °C per minute. Using a 10 µL syringe, add 1 µL of the ketone to 100 µL of methanol in a 1 dram vial, followed by a tiny granule (not more than a few milligrams) of 10-40 mesh sodium borohydride and swirl the liquid until the solid dissolves. Add about 0.5 mL of saturated aqueous sodium bicarbonate and about 1 mL of hexanes, cap the vial and shake it vigorously for a few seconds. Allow the layers to separate (the organic phase should be clear). Then use a clean 10 µL GC syringe to take about 1 µL of the upper phase (not the water layer!) for GC analysis using the same conditions as above (Note 2). Use the ketone and alcohol isomer ratios to determine the stereochemistry of the major and minor ketone isomers. In your notebook include photocopies (reduced as necessary) of all chromatograms and be sure to clearly explain your reasoning, giving structures of the possible products.

Notes

1. Use a ketone concentration of 1 μL in about 1 mL of dichloromethane and a 1 μL injection. The instrument should be equipped with a 0.2-0.32 mm ID HP-5 (SE-54) capillary column with split injection at a split ratio of approximately 40:1. Under these conditions, all of the compounds are baseline resolved. If necessary, the initial oven temperature can be lowered 5-10 °C to improve the separation.

2. If the GC shows some unreacted ketone remaining (identified by retention times), you must take this into account when evaluating the relationship between products and reactants. Do not assume that each isomer of the ketone reacts at the same rate.

Experiment 2: Thin Layer Chromatography
A. TLC of Compounds Containing a Variety of Functional Groups

TLC is a quick, powerful and inexpensive means of analysis. The main uses of TLC are to

(a) monitor reactions to ascertain completeness (absence of starting material).

(b) check purity of compounds.

(c) determine proper solvent for column chromatography.

(d) analyze fractions from column chromatography.

It is helpful to have some idea of how various functional groups behave under TLC conditions (i.e., what R_f values are observed in what solvents). In this experiment, you will perform TLC analysis on representative members of the most common classes of compounds. Your instructor will assign specific compounds

Type of Compound	Compound*
hydrocarbon	
ether	
ketone	
aldehyde	
✓ ester	
✓ alcohol	
carboxylic acid	
✓ amine	

* To be specified by your instructor.

By carrying out TLC of these compounds in several solvents, you will get a good feel for their relative polarities. In future experiments, this experience will allow you to guess what solvents are appropriate for analysis of various compounds. You will also learn how easy (or difficult) it is to visualize various types of compounds on the TLC plate.

Before proceeding, be sure to read the discussion of thin layer chromatography (p. 51).

Procedure

First prepare several spotting capillaries as described in the TLC discussion (p. 51). You do not need a new capillary for each spot! They are easily cleaned by drawing up some acetone and draining it onto a paper towel a few times, but be sure they drain completely. The TLC plates are cut from 20 cm x 20 cm plastic-backed silica sheets using scissors. Since you will be spotting eight different compounds on each TLC plate, you must make the plates wider than usual. A width of 4-5 cm should be sufficient, and the height will be 6.7 cm (one-third of 20 cm). With a pencil (not pen!), make eight marks about 1-1.5 cm from the bottom and evenly separated from each other (see Figure 16). Keep the first and last marks at least 5 mm from the edge, because the solvent often runs slightly faster at the edges. Carefully write below each mark some notation which will remind you what compound to spot at that position. Then obtain a set of the solutions from your instructor; these will contain 2-5% of the various compounds dissolved in dichloromethane or hexanes. Spot the compounds in the appropriate positions, touching the capillary *quickly* once or twice to each spot. Small spots are better than big ones! Practice making them as small as possible as you proceed through this experiment. Clean the capillary with acetone between different compounds. Using a total of five separate plates, carry out the TLC analyses in the following solvents:

Solvent mixtures:

hexanes

5% ethyl acetate in hexanes

10% ethyl acetate in hexanes

20% ethyl acetate in hexanes

50% ethyl acetate in hexanes

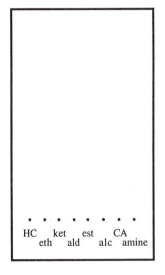

TLC plate shown approximately full size (6.7 cm high x 4 cm wide)

Figure 16: Thin layer chromatography plate shown approximately full size (6.7 cm high x 4-5 cm wide).

For the developing chamber, you will need an appropriately-sized jar with a tight-fitting cap. Place a piece of filter paper on the bottom to prevent the TLC plate from slipping. Then add the solvent you will be using (bottles of premixed solvents may be provided), *but do not add too much!* The spots **must** be above the solvent when you insert the TLC plate! Use forceps to insert and remove the plates. Because evaporation will quickly alter the ratio of hexanes to ethyl acetate, *keep the jar sealed as well as possible before and during the elution of the plate!* When the solvent has reached within a few millimeters of the top of the TLC plate (about 5 minutes), remove it and *immediately* mark the solvent front with a pencil. Allow the solvent to evaporate for about 30 seconds, then proceed with the three-stage development procedure below.

First, being careful not to look directly at the light, expose the plate to UV light and mark with a pencil whatever spots are visible. On the plate, somehow note which spots are weak, medium, or strongly UV active. Next, place the plate in a jar containing I_2/silica gel, letting it be covered with the mixture for 15 seconds or so. Upon removing it, note *immediately* which spots are visible (brownish on a white background). The iodine will evaporate quickly, and you should keep the plate in a hood. Finally, treat the plate with PMA solution[1] by either dipping it quickly or (preferably) spraying it with the solution, then dry it briefly and develop the spots by heating on a hot plate (~ 150 °C) for 30-60 seconds. You will be able to see the spots from behind (which sometimes shows things more clearly) if you wipe any excess PMA off of the back of the plate with a moist paper towel.

To change solvents in your TLC jar, pour the old solvent into an organic waste bottle and let the jar dry in the hood (removing the filter paper speeds this up) or use compressed air to complete the evaporation.

In your notebook, trace each plate in pen, but shade in the spots with a pencil. You must somehow note how different spots responded to each visualization method. You may wish to draw in the plate as it appears with PMA, then to the side note what spots were visible with UV and I_2 and how strongly. Then, <u>from the original plates</u>, measure the R_f values of each component on each plate. Tabulate the data and prepare a graph of R_f versus % ethyl acetate for each of the compounds (one graph with eight lines will do if the lines do not interfere with each other too much). Also, make conclusions as to what types of compounds are visualized by each of the methods you used.

1. The PMA reagent is a 5% solution of phosphomolybdic acid in ethanol.

Questions
*1. Do you think that different compounds with the **same** functional group would have the same R_f values or not?*

2. Which, if any, of the compounds you used were chiral? Were the enantiomers separated from each other by TLC?

3. Why is it better to use a pencil than a pen when you mark the starting positions?

4. Would there be any effect on the results if you had let the plate overdevelop (i.e., let the solvent reach the top and stand that way for a while)?

5. TLC plates which have been freshly dried in an oven will exhibit all of the spots at slightly lower positions; Why?

6. Why do you want the initial spots as small as possible? (two reasons)

Exploring Further...

1. Determine the response of the various functional groups to vanillin stain. This stain is prepared as described in the TLC discussion (p. 53), and is applied and developed by heating the same way the PMA stain is used.

2. Choose compounds which represent 1°, 2° and 3° alcohols and compare their behavior on TLC. Do they have the same R_f values or not?

B. A Multisolvent Study of a TLC Separation

When attempting difficult liquid chromatography separations, it is worthwhile to take the time to evaluate several mixtures of different solvents to determine which gives the best separation. This is done using TLC, and the efficiency of a given solvent mixture is determined by the observed ΔR_f's. However, for an accurate evaluation, the solvent mixture should be chosen such that the compounds appear in approximately the correct region of the plate ($R_f \sim$ 0.2-0.4).

Treatment of a mixture of guaiazulene and trimethylazulene with an insufficiency of trifluoroacetic anhydride produces a mixture of the compounds shown below. In this experiment, the response of the separation to various solvents will be explored. TLC will be used to determine which of several solvent mixtures gives the best separation of the compounds. The separation is measured by the difference in R_f values of adjacent spots.

Solvents: 40% toluene in hexanes
              20% dichloromethane in hexanes
              15% diethyl ether in hexanes
              10% ethyl acetate in hexanes

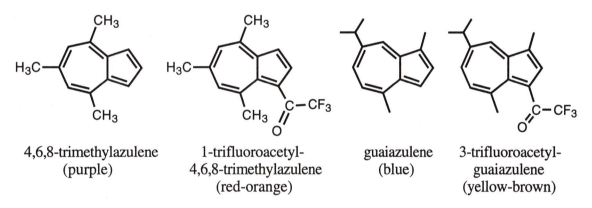

| 4,6,8-trimethylazulene (purple) | 1-trifluoroacetyl-4,6,8-trimethylazulene (red-orange) | guaiazulene (blue) | 3-trifluoroacetyl-guaiazulene (yellow-brown) |

Procedure

The mixture of compounds and the chromatography solvents will be provided. There is no need to do full visualization of the plates (UV, I$_2$, PMA) because the compounds are colored (but <u>do</u> expose one plate to I$_2$; record what effect it has on the spots). Record the TLC plates in your notebook promptly, as some of the spots will fade with time. Discuss which solvent mixture gave the best separation.

Questions

1. When determining the best solvent, does it depend on which spots one wishes to separate?

2. What order did the compounds elute in?

3. What effect did iodine have on the plate?

Experiment 3: Introduction to "Flash" Column Chromatography:
Column Chromatography of a Dye Mixture

Column chromatography is a moderately powerful technique for the separation of organic materials; although the separating power is limited (< 50 theoretical plates), like other forms of liquid chromatography it can be applied to most types of compounds. The only restrictions are that the compounds to be separated must (a) be soluble in the solvents being used and (b) have enough difference in polarity to cause them to be separated (i.e., travel at different rates) on the column. For these reasons, column chromatography is probably the most frequently used tool to purify organic materials on small-scale (say, two grams or less of product).

In this experiment, you will separate a two-component mixture using column chromatography. We will be using colored compounds (a) so that the separation will be obvious and (b) so that you can practice your technique (for example, Experiment 4 requires a <u>careful</u> column chromatography separation of <u>colorless</u> materials). The materials in the mixture are shown below. Can you guess in which order these will elute from the column?

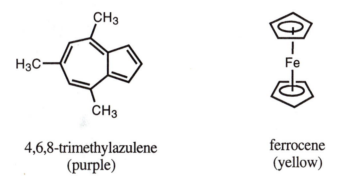

4,6,8-trimethylazulene
(purple)

ferrocene
(yellow)

Note: Please be careful with the chromatography columns; they are expensive and difficult to replace!

Procedure:

Carefully read the material on column chromatography (p. 54); the details on how to set up and run a column are given there. Prepare a chromatography column using 30 g (60 mL) of silica gel as a slurry in hexanes, and apply a thin layer of sand to the top. Apply 2 mL of the chromatography solution provided, which is a hexanes solution of trimethylazulene and ferrocene (10 mg and 50 mg, respectively). After rinsing the compounds carefully into the silica, add 200 mL of hexanes and elute the column with hexanes until the yellow and purple bands have been eluted. Collect these bands in Erlenmeyer flasks (125 or 250 mL). *During development of the column, be sure to reuse any colorless solvent which elutes, including that which you used to pack the column*. Using a graduated cylinder, measure the volume in which each band was collected and record these values in your notebook. At your instructor's discretion, you may recycle the solvents by simple distillation.

Cleaning the column: When all of the solvent has eluted (i.e., column appears dry), disconnect it from the gas supply and connect the bottom to your water aspirator hose. Use the aspirator to pull air through the column for 10-20 minutes, or until the silica is dry enough to be forced out easily with a small amount of gas pressure. Discard used silica only in a "silica gel waste" bottle, not in the wastebasket. Rinse the column (if necessary) with water followed by acetone and return it to the storage area.

Experiment 4: Separation and Identification of Diastereomers

A. Reduction of 4-*tert*-Butylcyclohexanone

The reduction of ketones (and aldehydes) to alcohols is easily accomplished by treatment with sodium borohydride, using methanol as solvent. Although sodium borohydride is a hydride ("H⁻") donor, it does not react rapidly with water or alcohol solvents in the absence of acid. More powerful hydride donors, such as lithium aluminum hydride (LiAlH₄), react instantly and violently with water and alcohols.

As shown below, the reaction of ketones with sodium borohydride first produces a "borate ester", in which the alcohol-to-be is attached to boron. Because these B-OR bonds are easily hydrolyzed (exchanged with water or alcohols), the methanol solvent displaces the product from boron.

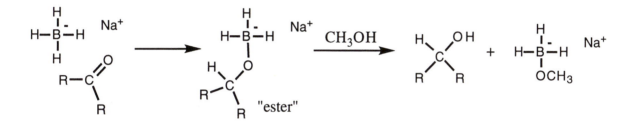

In the case of cyclohexanones, there are two possible directions (axial and equatorial) from which the hydride can attack the carbonyl group. For substituted cyclohexanones, such as 4-*tert*-butylcyclohexanone, this leads to two possible products (cis and trans), and these are **not** formed in equal amounts. The cis and trans products are geometrical isomers (one type of diastereomers) with different properties, NMR spectra, etc., and these differences can be used to both separate and identify them. Separation will be accomplished by careful column chromatography, since the isomers have somewhat different R_f values on silica gel. Since the cyclohexane ring contains a large substituent (*tert*-butyl), the ring will be "locked" into a single chair conformation because placing this group in an axial position is very unfavorable energetically. In cases like this, is it usually possible to identify the cis and trans isomers using proton NMR by evaluation of the chemical shift and coupling constants exhibited by the **H**-C-O proton, which will be permanently either axial or equatorial. Refer to an intermediate spectroscopy text for details.

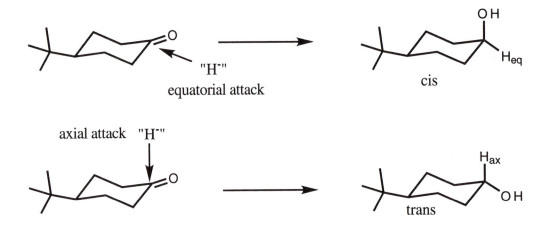

equatorial attack

cis

axial attack   "H⁻"

trans

Procedure

Preparation of alcohols: To a test tube with a small stir bar ("flea") is added 4-*tert*-butylcyclohexanone (300 mg) and 1 mL of dry methanol. Use magnetic stirring to dissolve the ketone, then stopper the tube and cool it to 0 °C. Then *cautiously* add fresh granular $NaBH_4$ (30-35 mg, 10-40 mesh) with stirring (may be exothermic). [*Note:* If powdered borohydride is used, add it carefully in 2 or 3 portions, using extra caution to avoid the possibility of having the mixture foam uncontrollably.] The solution is then allowed to stir for 5 minutes, or until the borohydride dissolves. Then carefully acidify the mixture (may foam!) with 1 M HCl (2 mL), followed by extracting twice with 2 mL portions of hexanes. Each hexanes extract is carefully pipetted away from the water layer (Figure 17) and passed through a 2 cm column of anhydrous granular

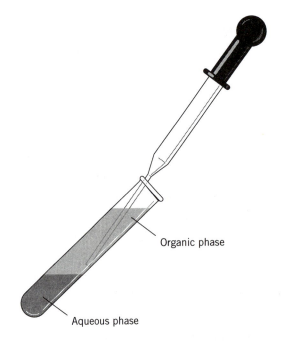

Organic phase

Aqueous phase

Figure 17: Removing organic phase (hexanes) by pipette.

Na_2SO_4 packed in a disposable pipette; the filtrate is collected in a tared round bottom flask. A sample is then taken for capillary GC analysis (dilute one drop with 1 mL of hexanes; temperature program will be 100 °C to 150 °C at 5 °C per minute). At your instructor's discretion, determine (by TLC) what mixture of ethyl acetate in hexanes is appropriate for a column separation, but before doing the column, check your choice of solvent with your instructor. Then remove the solvent by rotary evaporation. *Try to have all of these steps complete the first day so that you will be ready to start the column at the beginning of the second lab period.*

Purification: Separate the diastereomeric alcohols by column chromatography on 30 g of silica using a mixture of ethyl acetate in hexanes (see discussion of column chromatography, p. 54, for details). Since your crude material is a solid, you should dissolve it in a minimum amount (1-2 mL) of hexanes for application to the column, and rinse the flask and transfer pipette once or twice with <u>small</u> amounts of hexanes.

Before applying your compounds to the column:
Make up 150 mL of the solvent you choose *in addition to* what you need to pack the column, plus 100 mL of a mixture containing 5% more ethyl acetate. When the first 150 mL runs out, continue the elution with the more polar solvent (without letting air into the column between solvents). Be prepared to collect up to 40 fractions (see column chromatography discussion (p. 56) as to what size fraction is appropriate). You should be able to collect at least 50 mL of forerun.

Note: Be especially careful when applying the compounds to the column; volatile organic solvents in a pipette have a tendency to squirt out unexpectedly. Rinsing the pipette with the solvent beforehand will serve to equilibrate it with solvent vapor and greatly reduce this tendency.

Analyze your fractions by TLC; because the solutions are now rather dilute, you should <u>spot each fraction at least 10 times</u>. (*Hint:* You can usually get an idea of roughly which tubes contain your compounds because after standing in the hood for a few minutes there will often be a ring of residue just above the solvent level.) Once you have located the compounds, combine in separate *tared* flasks those fractions containing each pure diastereomer. There may be several "mixed fractions" which contain some of both isomers; combine these also in a different flask.

This is a good place to stop; store the flasks well sealed and upright!

Before concentrating the combined fractions, prepare <u>three</u> samples for capillary GC of (a) your pure major diastereomer, (b) your pure minor diastereomer, and (c) your combined mixed fractions (if any). Make the GC vials about $1/3$ full with your sample <u>but do not dilute them with additional solvent</u>, and carry out the GC analysis. Then concentrate the major and minor diastereomer solutions by rotary evaporation. Once the solvent appears to be gone, bring the flasks to constant weight using aspirator vacuum (record the weight after perhaps every 2 minutes of vacuum exposure until it does not change significantly for three weighings; this data should be in the notebook, preferably in graph form). Determine the weight yield and percent yield of each isomer. (*Note:* The melting points are not very different, and therefore not of much value to obtain.) At your instructor's discretion, obtain proton NMR spectra for each pure compound. Analyze the NMR spectra to determine which isomer (major or minor) is cis and which is trans (explain/show your analysis of the spectra in your notebook). Using this information, note the identity of your capillary GC peaks on the chromatograms. All TLC plates should be drawn in, and all chromatograms (with peaks labeled and temperature program noted) and NMR spectra should be reduced appropriately and taped into your notebook in the appropriate places.

Questions

1. *How would the following affect purification by column chromatography (improve the separation, decrease the separation, or have no effect)? Why?*

  a. *You use a solvent which gives a TLC R_f of 0.55 for the compound you want.*

  b. *You add the silica dry to the column instead of as a slurry.*

  c. *Compound is put on as a dilute solution rather than as a concentrated solution.*

  d. *You drop the pipette into the top of the column.*

  e. *You load the compounds onto the column then take a 30 minute break.*

  f. *You collect fractions equal (in mL) to the amount of silica you used (in grams).*

  g. *When you analyze the fractions by TLC, you spot each one only once to the plate.*

2. *Based on the amounts you used in this experiment, how much silica and solvent would be required to purify 4.2 g of a mixture of cis- and trans-4-tert-butylcyclohexanol?*

3. *If you had to separate a mixture containing equal amounts of two materials with very similar polarities, do you think you would get more of the top spot or of the bottom spot pure? (Hint: tailing is common in flash chromatography; see Figure 12, p. 42.)*

Exploring Further...

1. If time and instrumentation allow, and with your instructor's permission, analyze each fraction separately by GC. If you choose to do this, try to avoid evaporation of the fractions prior to analysis. Inject the same amount of each fraction and plot the area of each diastereomer versus fraction number (or mL). In this way, you can construct a crude "chromatogram" which

corresponds roughly to what the output of a liquid chromatography detector would have given if the flash column had been so equipped. Although you needn't analyze the forerun by TLC, include the fraction equivalent (or mL) of it as a "blank baseline" in order to establish the "retention time" (or volume) of the peaks.. You can then estimate the separating power (in theoretical plates) of the column using the formula below, where the retention time and peak width must have the same units. This formula applies to any type of chromatography where a symmetrical peak of defined width at half-height is obtained (e.g., GC, HPLC).

$$\text{number of theoretical plates} = 5.54 \left(\frac{\text{retention time}}{\text{peak width at half-height}}\right)^2 \qquad \text{(Equation 5)}$$

2. Carry out the reaction using a very large hydride donor such as lithium tri(*sec*-butyl)borohydride (L-Selectride®, available from Aldrich). How does the diastereomer ratio change?

3. On a $^1/_{10}$th scale, carry out the reaction at different temperatures to see how the isomer ratio is affected. For example, with the ketone in boiling methanol (~ 68 °C), cautiously add sodium borohydride <u>very slowly</u> (watch for foaming!), then cool and work up as above. Alternatively, dilute the ketone with enough methanol to keep it soluble at dry ice temperature (-78 °C) and treat it with sodium borohydride. In this case, the reaction may require considerable time, so you must be prepared to maintain the low temperature until the reaction is complete (as determined by TLC?). See if the diastereomer ratios at a given temperature can be predicted by the equation

$$\Delta E_a = RT \ln \frac{k_1}{k_2} \qquad \text{(Equation 6)}$$

where

ΔE_a = difference in activation energies (calculated using ratio observed at any defined
   temperature)

R = gas constant

T = absolute temperature at which reaction occurs (K)

$\dfrac{k_1}{k_2}$ = ratio of major and minor diastereomers

To determine percent major diastereomer from the ratio, use the relation

$$\% \text{ major diastereomer} = \frac{k_1/k_2}{(k_1/k_2) + 1} \qquad \text{(Equation 7)}$$

This assumes ΔE_a is invariant with temperature, and that you know the temperatures at which each reaction occurs with some accuracy.

B. *cis*- and *trans*-1-(4-Fluorophenyl)-4-*tert*-butylcyclohexanol: An Organic Compound with an Unusual Property

Addition of Grignard reagents (RMgX) to carbonyl compounds is an excellent method for the preparation of alcohols. The immediate product of the reaction is the halomagnesium salt of the alcohol, which hydrolyzes in the presence of water to give the free alcohol and a halomagnesium hydroxide. This form of the magnesium salt is a troublesome precipitate, and often acid is added to dissolve the solid. When preparing acid-sensitive organics, ammonium chloride solution serves as a mild acid for this purpose.

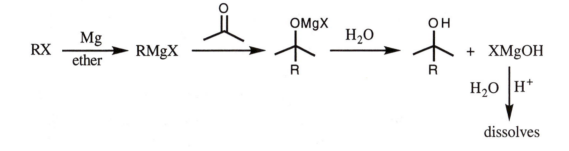

Because Grignard reagents are air and water sensitive, the preparation requires the use of dry glassware and solvents, and preferably inert atmosphere. The halide can be chlorine, bromine or iodine; however, fluorine is too unreactive. The reaction also requires the use of ether solvents, which stabilize the Grignard reagent by coordinating to the Lewis-acidic magnesium atom.

In most cases, Grignard reactions proceed in high yields, as in this experiment. However, there are certain side reactions which are sometimes of importance. Rather than adding to the carbonyl group, Grignard reagents (and organolithiums) can sometimes do either or both of the following:

(a) Enolization: The Grignard can act as a base to enolize the aldehyde or ketone. The enolate reverts to the carbonyl compound upon water workup, and can make it appear as if an insufficient amount of the Grignard reagent was added. This reaction is most important when one attempts to add hindered (2° or esp. 3°) Grignard reagents to hindered ketones. For example, the reaction of methylmagnesium iodide with camphor (a hindered enolizable ketone) results in about 60% enolization and only 40% addition.

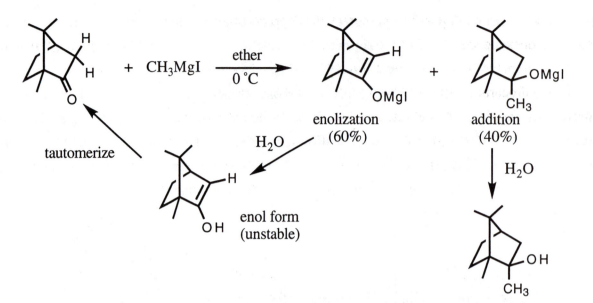

(b) Reduction: If the Grignard reagent has hydrogens β to the metal, a reduction reaction is possible. For example, attempted reaction of *tert*-butylmagnesium bromide with di-*tert*-butyl ketone results in almost complete reduction (see F. C. Whitmore, R. S. George *J. Am. Chem. Soc.* **1942**, *64*, 1239). Note that certain Grignard reagents (methyl, phenyl) have no β hydrogens or are otherwise incapable of this reduction reaction.

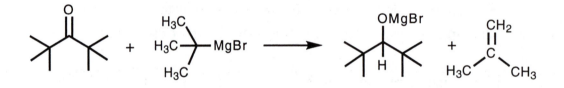

Finally, the formation of a common by-product occurs during the preparation of the Grignard reagent itself. Dimerization of the alkyl or aryl groups during Grignard reagent formation can occur, shown below for the reaction of bromobenzene and magnesium. This is generally a minor side reaction, but the coupling product can usually be detected in the crude product if a sufficiently sensitive analysis is used (e.g., capillary GC).

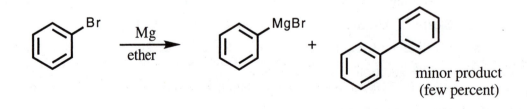

In this experiment, you will prepare 4-fluorophenylmagnesium bromide and add it to *4-tert*-butylcyclohexanone. A mixture of cis and trans isomers is obtained; the naming here refers to the relationship between the *tert*-butyl group and the OH group. Note that in the trans isomer, one of the two hydrocarbon groups will be axial if a chair conformation is present. The *tert*-butyl group is very bulky and energetically unfavorable for placement in an axial position. The phenyl ring is somewhat less bulky, and it is at least 1.1 kcal/mol easier to place in an axial position than is the *tert*-butyl group. This has the effect of forcing the aromatic group in the trans isomer to be in an axial position, while the cis isomer has both groups in equatorial positions.

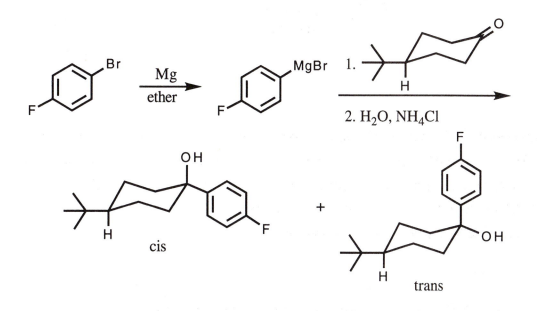

The reaction produces both isomers in significant amounts. The diastereomers are separable by column chromatography, but because there is no **H-C-OH** with which to assign axial/equatorial coupling constants and chemical shifts, assignment of the stereochemistry is not as easy as in Experiment 4A. It is left to the student to determine the relative stereochemistry using a literature reference (E. W. Garbisch, D. B. Patterson *J. Org. Chem.* **1963**, *85*, 3228; beware of opposite use of the terms cis and trans!; and: G. D. Meakins, R. K. Percy, E. E. Richards, N. R. Young *J. Chem. Soc. Sec. C* **1968**, 1106). A further interesting aspect of this reaction is that one of the products has a very unusual property that you will inevitably discover during this experiment.

Caution: 4-Fluorobromobenzene is toxic; wear gloves and handle the material (and rinse the syringe) only in the hood. Diethyl ether is very flammable: do not allow flame or spark sources in the vicinity of ether in open containers.

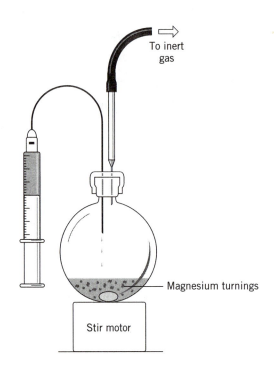

Figure 18: Reaction setup for Grignard experiment.

Procedure

Weigh approximately 220 mg of magnesium turnings into an oven-dried 50 mL flask and add a magnetic stir bar. Seal the flask with a septum and place it under passive inert atmosphere (see Figure 18). Add 5 mL of anhydrous ether by syringe, and stir the suspension while 4-fluorobromobenzene (1.0 mL) is added cautiously as follows: add about one-fifth of the halide, then wait (up to 10-20 minutes if necessary) for signs of reaction occurring (clouding) before slowly adding the remainder over 5-10 minutes. Avoid adding the halide so rapidly that reflux occurs. Let the mixture stir for 1 hour, during which time the amount of magnesium should noticeably decrease. At this point, cool the reaction flask to 0 °C. Then dissolve 4-*tert*-butylcyclohexanone (1.1 g) in 10 mL of dry ether and add the solution *slowly* to the Grignard reagent by syringe or cannula. After stirring for 15 minutes (longer is OK), add a few drops of water *cautiously*. When any exothermicity is no longer evident, add 10% ammonium chloride solution (20 mL). Then transfer the mixture to a separatory funnel using 20 mL of additional ether, separate the aqueous phase, and wash the organic phase once with saturated sodium chloride solution ("brine"). After drying the organic phase over anhydrous magnesium sulfate, prepare a sample for capillary GC (180 °C to 230 °C at 5 °C per minute) by diluting a 2-4 drops with 1 mL of dichloromethane in a GC vial. There should be little or no ketone evident (~ 2.7 minutes) in the GC chromatogram. Then remove the solvent by rotary evaporation, applying aspirator vacuum as necessary until approximately constant weight is obtained; this may take some time.

The diastereomers may be separated by column chromatography (see p. 54) on silica gel using 100% CH_2Cl_2 (use this relatively toxic solvent only in a hood). You will need to dissolve the crude product in a minimum volume of dichloromethane to load it onto the column; this will require 5-10 mL per gram of solid. Do the separation using 35-40 g of silica per gram of crude product, and the separation will require about 300 mL of solvent per gram of crude product (not including what is required to pack the column). Thus, whether you chromatograph your entire product, or only a portion, will depend on the size of columns you have available. Be aware that the products move rather quickly in dichloromethane (R_f ~ 0.6 and 0.4), so do not collect very much forerun.

Note: Be especially careful when applying the compounds to the column; volatile organic solvents in a pipette have a tendency to squirt out unexpectedly. Rinsing the pipette with the solvent beforehand will serve to equilibrate it with solvent vapor and greatly reduce this tendency.

Analyze your fractions by TLC; because the solutions are now rather dilute, you should spot each fraction at least 10 times. (*Hint:* You can usually get an idea of roughly which tubes contain your compounds because after standing in the hood for a few minutes there will often be a ring of residue just above the solvent level.) Once you have located the compounds, combine in separate *tared* flasks those fractions containing each pure diastereomer. There may be several "mixed fractions" which contain some of both isomers; combine these also in a different flask.

Before concentrating the combined fractions, prepare up to three samples for capillary GC of (a) your pure major diastereomer, (b) your pure minor diastereomer, and (c) your combined mixed fractions (if any). Prepare the GC samples placing ~ $1/3$ mL of each sample in separate GC vials but do not dilute with additional solvent, and carry out the GC analysis (180 °C to 230 °C at 5 °C per minute). Then concentrate the major and minor diastereomer solutions by rotary evaporation. Once the solvent appears to be gone, bring the flasks to constant weight using your hood aspirator (record the weight after perhaps every two minutes of vacuum exposure until it does not change significantly for three weighings; this data should be in the notebook, preferably in graph form). Determine the weight yield and percent yield of each isomer.

After isolating each isomer in pure form, at your instructor's discretion, obtain ^1H, ^{13}C and ^{19}F NMR on each; use $CDCl_3$ solvent, but do not use more than ~ 80 mg of compound in each

NMR tube. Also, dissolve about 20-40 mg of each pure diastereomer in separate vials containing 1-2 mL of hexanes; heating will be necessary to accomplish the dissolution. Then let the vials stand at room temperature for 10-15 minutes; if necessary, cool briefly in ice water. Record the observations in your notebook.

Questions

1. Given the structure of the desired products, why do you suppose that a mineral acid (HCl or H$_2$SO$_4$) was not used to dissolve the magnesium alkoxide salt during the workup?

2. One of the most easily recognized patterns in proton NMR spectra is that of a para-disubstituted aromatic ring substituted with two different groups. Do the compounds prepared in this experiment show the same pattern? Explain the pattern observed in the aromatic region of the proton NMR spectrum. If you obtained the ^{19}F NMR spectrum, explain this as well.

3. How many carbon resonances do you expect for a para-disubstituted aromatic ring with two different groups? If you obtained the ^{13}C NMR spectrum, is this what you observed? Explain. Can you assign which signals belong to which aromatic carbons? On what basis?

Experiment 5: NMR-Related Procedures

See also

    Experiment 4B: 1-(4-Fluorophenyl)-4-*tert*-butylcyclohexanol

                (describes the preparation of a molecule containing the NMR-active ^{19}F)

    Experiment 10C: Optical purity analysis of 1,2-diphenyl-1,2-ethanediol

                (describes use of ^{31}P NMR to determine diastereomer ratios)

A. Kinetics of Dimerization of Cyclopentadiene

Cyclopentadiene is useful as a diene in Diels-Alder reactions and is also the precursor to the cyclopentadienyl ligand which is common in organometallic chemistry. Cyclopentadiene is not commercially available because, upon standing, it dimerizes via a Diels-Alder reaction to give dicyclopentadiene, which is sold commercially. Monomeric cyclopentadiene is prepared by "cracking" (fragmenting) the dimer at 180-200 °C. This process is simply a reverse Diels-Alder reaction. Fortunately, cyclopentadiene distills at about 40 °C and is easily separated from the dimer (bp 170 °C) by distillation.

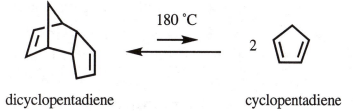

              dicyclopentadiene   cyclopentadiene

It is important to know the rate at which cyclopentadiene dimerizes because this determines how quickly one must use it after cracking. The dimerization reaction can be monitored either by proton NMR or by gas chromatography. However, gas chromatography is more difficult because the extreme volatility of cyclopentadiene requires use of a high-boiling GC solvent, and because the detector's response to each of the compounds must be calibrated. In this experiment you will use proton NMR to quantitatively monitor the progress of the dimerization.

Caution: Cyclopentadiene and dicyclopentadiene have strong irritating odors. Keep glassware containing either of these in the hood. Excess material should be returned to the dicyclopentadiene bottle. Allow cyclopentadiene residues to evaporate in a hood (occurs quickly), and rinse dicyclopentadiene residues into a waste bottle kept in the hood.

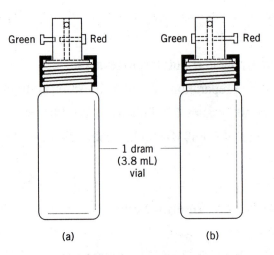

Figure 19: Mininert®-sealed vial in (a) closed and (b) open positions.

Procedure

Preparation of cyclopentadiene: Place in a 25 mL round bottom flask 10 mL of dicyclopentadiene and a stir bar. [**Note:** Old samples are best passed through a small amount of alumina to remove any peroxides.] Attach the flask to a vertical condenser packed loosely with stainless steel mesh, and the assemble the remaining apparatus for distillation; the final result should be identical to Figure 7a, except there will be an extra condenser (which acts as a fractionating column) between the flask and the distillation adapter. Connect the apparatus to passive inert atmosphere or to a calcium chloride drying tube. *Pass water only through the (approximately) horizontal condenser jacket, not the vertical one!* With good magnetic stirring, heat the flask until distillation commences; this will require a bath temperature of at least 180-200 °C. The distillation should proceed with a head temperature of about 40 °C; if the temperature should rise more than a few degrees above this value, immediately slow the distillation rate by reducing the amount of heat being applied. Keep the receiver flask in ice. When 2-3 mL has been collected, stop the distillation. If the distillate is cloudy, pass it through a 2 cm column of anhydrous sodium sulfate. Return the undistilled material which remains to the original bottle. (See cautions above regarding odor control during cleanup.)

Reaction setup and sampling: Transfer approximately 1 mL of cyclopentadiene to a 1 mL conical vial equipped with a Mininert® cap (see Figure 19) and note the time; this will be considered the start of the reaction. If possible, the vial should be kept at a constant temperature, at least approximately. Wait at least two hours before taking the first sample; otherwise there will not be enough dimer present to accurately measure. You should sample the reaction at least twice each day for the first 2-3 days, with intervals of at least several hours between samples. At appropriate intervals, sample the reaction by removing approximately 20 μL by microliter syringe and immediately adding this to approximately 0.4 mL of CDCl$_3$ in an NMR tube. Cap the tube quickly to avoid the loss of the very volatile cyclopentadiene to evaporation, then shake the tube briefly to mix the contents. Carefully note the time at which each sample is taken. You should either obtain the proton NMR spectrum immediately or store the sample in a freezer until the analysis. The ratio of monomer to dimer is determined from the spectrum as described below.

Alternatively, an NMR tube can be used as the reaction vessel, which avoids the need for a Mininert®-capped vial. If you choose this approach, place 0.6 mL of cyclopentadiene in an NMR tube and line the interior of the plastic cap with aluminum foil to help prevent evaporation of the volatile monomer. Because no deuterated solvent is present, you must run the NMR spectrum unlocked. Try to keep the tube at constant temperature when not running an analysis. This method does not allow you the luxury of saving samples for later analysis.

Analysis of the NMR spectra: Monomeric cyclopentadiene exhibits vinyl resonances as two complex multiplets at 6.45 and 6.55 ppm. The dimer has vinyl resonances at 5.5 and 5.95 ppm. When you obtain the NMR spectrum, you should carefully integrate these regions of the spectrum. If necessary, you should adjust the phasing of each integral so that it is as level as possible before and after the peak. Because each compound has a total of four vinyl resonances, and two monomers are required to form one dimer, the following expression relates the NMR peak areas to the percent of monomer remaining:

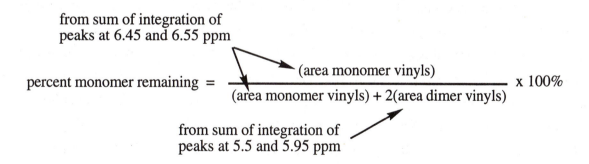

$$\text{percent monomer remaining} = \frac{\text{(area monomer vinyls)}}{\text{(area monomer vinyls)} + 2\text{(area dimer vinyls)}} \times 100\%$$

from sum of integration of peaks at 6.45 and 6.55 ppm

from sum of integration of peaks at 5.5 and 5.95 ppm

The Diels-Alder reaction typically proceeds by a concerted mechanism, and reaction is expected to exhibit second-order behavior. The rate of the reaction should be described by an equation of the form:

$$\text{rate} = \frac{d[\text{monomer}]}{dt} = k\,[\text{monomer}]^2 \qquad \text{(Equation 8)}$$

where: k = second order rate constant for the reaction

The kinetics of the dimerization are quantified by first rearranging and integrating the above expression to give:

$$\frac{1}{M_t} - \frac{1}{M_0} = kt \qquad \text{(Equation 9)}$$

where:

M_0 = initial monomer concentration

M_t = monomer concentration at any given point during the reaction

which is the same as:

$$\frac{1}{M_t} = kt + \frac{1}{M_0} \qquad \text{(Equation 10)}$$

Examination of the above expression suggests that plotting $\frac{1}{M_t}$ versus time will yield a straight line with a slope equal to the rate constant k and an intercept equal to $\frac{1}{M_0}$. Because we are mainly interested in the slope of the line, we will plot $\frac{1}{\% \ monomer}$ versus time rather than actual concentrations. The practical value of knowing the rate constant is that one can determine the half-life of the reaction. Because after one half-life ($t_{1/2}$), $M_t = \frac{M_0}{2}$, this can be substituted into the above expression to give:

$$t_{1/2} = \frac{1}{kM_0} \qquad \text{(Equation 11)}$$

Note that the half-life of a second-order reaction is dependent on the initial concentration of the reactant(s), unlike a first-order reaction. In your notebook, provide graphs of (a) percent monomer vs time, and (b) $\frac{1}{percent \ monomer}$ vs time, with the equation of the best straight-line fit from a linear regression analysis also provided. Calculate the half-life of the dimerization reaction; remember to use units of percent monomer remaining when inserting the value of M_0 into the half-life expression.

Questions

*1. Why is it **not** necessary to analyze samples immediately after dilution with CDCl₃?*

2. Do you see evidence in the ¹H NMR spectra of the dimer that both endo and exo isomers are formed?

Exploring Further...

1. Determine whether your data fits first order or second order kinetics better by comparing the linearity of each plot. First order reactions are evaluated by plotting $\ln M_t$ vs. time, and give a slope of -k and an intercept of $\ln(M_0)$. The half life is given by $\frac{\ln 2}{k}$.

2. You can determine the dimerization kinetics at lower temperatures by keeping the vial in a refrigerator or freezer. You should allow longer intervals between samples (perhaps one each day) and precool the syringe to avoid warming the sample prior to dilution in $CDCl_3$.

3. Determining the kinetics of the reaction by GC is possible, but much more challenging. If you attempt this, keep in mind the following:

(a) If analyzed by capillary GC, the samples must be diluted with a high-boiling solvent (dodecane) so that the early-eluting monomer not be obscured by solvent peaks. A temperature program of 80 °C to 140 °C at 5 °C per minute is suggested. (It may be possible to inject samples without dilution if a packed-column GC instrument is used.)

(b) The relative response of the detector to monomer and dimer must be determined. This is done by analyzing solutions containing known weights of pure monomer (obtained by cracking/distillation) and pure dimer (obtained by distillation at full aspirator vacuum; this lowers the boiling point sufficiently to avoid cracking). The pure materials are measured by weight to give a solution (in dodecane) of known "% monomer".

(c) The injection port temperature must be low enough to avoid the cracking reaction when the sample is injected. This can be determined by injecting samples of pure dimer at various injection port temperatures until you identify the highest temperature at which no monomer peak is observed. Interestingly, it has been observed that an injection port temperature of 200 °C gave negligible amounts (< 1%) of monomer, and 250 °C gave only about 5% monomer. Apparently the time in the injection region is short enough that the cracking reaction is easily avoided.

¹H NMR of cyclopentadiene (360 MHz).

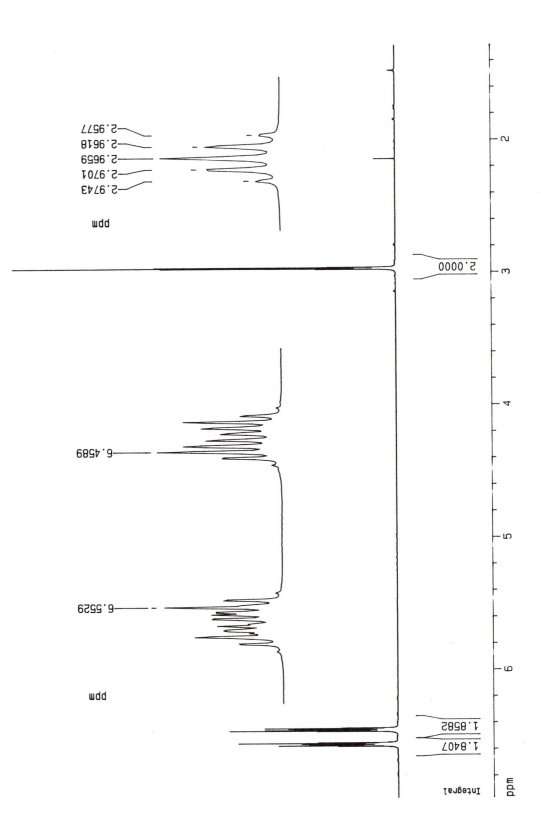

¹H NMR of dicyclopentadiene (360 MHz).

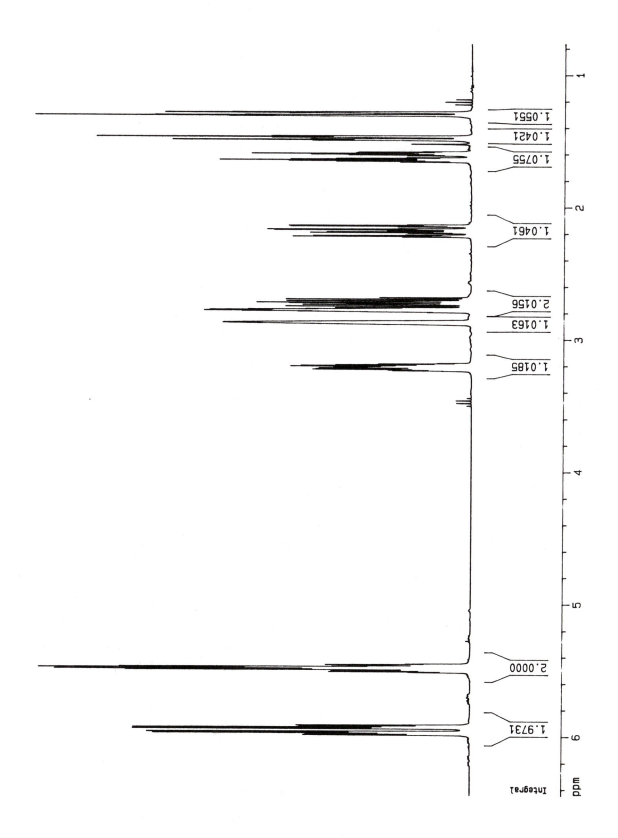

B. NMR Analysis of an Unknown

In this experiment, you will be given an unknown organic compound and the molecular formula. You are to prepare an NMR sample, obtain ^{1}H and ^{13}C NMR spectra and determine the structure of the unknown compound.

Preparing the sample: If necessary, see the discussion on cleaning and drying NMR tubes (p. 26). In order to obtain ^{13}C spectra relatively quickly yet have a reasonable lock signal strength, samples which are 20-30% compound and 80-70% solvent by volume are best. For solids, use 80-100 mg. If a superconducting ($\geq$ 200 MHz) spectrometer is available, somewhat less compound (10-20%) should be sufficient. The solvent will be $CDCl_3$ unless otherwise specified. The total volume of the NMR sample should not exceed 0.6 mL. The final sample should be clear and free of all solids; filter through a pipette with a small cotton plug into a clean NMR tube if necessary.

Caution: $CDCl_3$ is toxic to the liver and a carcinogen. Handle this material in the hood. It is also relatively expensive. Be very careful not to contaminate the bottle. It is a good idea to attach a test tube to the bottle to hold a disposable pipette, which helps minimize the risk of contamination. If you should accidentally contaminate the pipette by touching the compound, immediately discard that pipette and replace it with a new one.

Obtaining the spectra: It's usually best to do the proton spectrum first, because this allows you to verify that the peakshape and sample concentration is reasonably good before obtaining the carbon spectrum with its much greater sensitivity demands. Once you have obtained the proton spectrum, you should be able to tune the instrument to carbon without removing the sample: this allows you to maintain lock and shim settings, and saves time.

The proton spectrum: In addition to printing the spectrum, be sure to include
  (a) chemical shift marks, labeled
  (b) integral values shown (with proper phasing)
  (c) any recognizable patterns (e.g., dd or qd, etc.) expanded as necessary to show the number of peaks and allow measurement of coupling constants where possible.

There is usually no good reason to show extensive baseline regions without peaks; expand the portion of the spectrum which has information to fill the plot (but do not leave significant peaks out).

The carbon spectrum: In addition to the spectrum, include

(a) chemical shift marks, labeled

(b) chemical shifts of each peak.

In your notebook, you must detail how you used the spectra to determine the structure of the unknown compound. If ambiguity remains, specify within what limits the structure is known. You should consult tables of chemical shifts, coupling constants, etc., as an aid in solving the structure.

Experiment 6: The Diels-Alder Reaction

In 1928 Otto Diels and Kurt Alder reported that the reaction between 1,3-dienes and certain alkenes yielded cyclohexene derivatives. This became known as the Diels-Alder reaction, for which the discoverers were awarded the Nobel Prize in 1950. In the context of this reaction, the alkene is referred to as the dienophile. The Diels-Alder reaction is probably the most useful method for the construction of cyclohexene ring systems. The reaction is very difficult with simple dienes and alkenes, requiring high temperatures and long reaction times. However, electron donating groups in the diene or electron-withdrawing groups in the dienophile favor the reaction. In these cases, the reaction proceeds faster and/or at lower temperatures. In general, either the diene or the dienophile must be of the especially reactive type for the reaction to proceed easily.

The Diels-Alder reaction between isoprene and methyl acrylate is the subject of this experiment. Although isoprene is not an especially reactive diene, methyl acrylate is a good dienophile and the reaction can proceed at reasonable temperatures (130-140 °C, 24 h). It has been observed that certain Lewis acids can catalyze the Diels-Alder reaction, allowing it to proceed faster and/or at much lower temperatures than otherwise possible. However, not all Lewis acids are useful because some cause polymerization of the diene and/or the dienophile. In this experiment, aluminum chloride is used to rapidly accomplish a reaction at room temperature which would otherwise require extended heating. The Lewis-acidic aluminum complexes the carbonyl oxygen of the methyl acrylate, greatly increasing its reactivity.

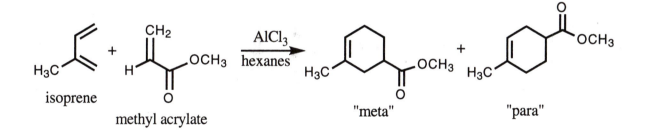

In this reaction, two products are possible, generically referred to as "meta" and "para" products (methyl 3-methyl-3-cyclohexenecarboxylate and methyl 4-methyl-3-cyclohexenecarboxylate, respectively). These isomers are not formed in equal amounts, and are very difficult to separate preparatively, but are separable by capillary GC. The second part of this experiment is to determine which isomer is the major product. This is accomplished using a transfer hydrogenation/dehydrogenation process involving three molecules of the cyclohexene product. In this reaction, one molecule of the product transfers hydrogen to two other molecules, yielding one aromatic and two saturated molecules. Because catalysts lower the activation energy for both the forward and reverse reactions, it should not surprise you that materials which catalyze hydrogenation can also catalyze dehydrogenations. In the reaction shown below, "Pd-C" refers to palladium metal dispersed on activated carbon. Interestingly, the addition of cyclohexene does *not* shift the reaction toward either the aromatic or saturated products (by either absorbing or releasing hydrogen, respectively); its presence simply accelerates the reaction, an unexpected observation made during the development of this experiment. The identity of the major Diels-Alder product can be determined by examination of the aromatic region of either the proton or carbon NMR spectra of the mixture following the palladium-catalyzed reaction. Prior to completing the second part of the experiment, consider what you would expect the NMR spectrum of each possible aromatic to look like.

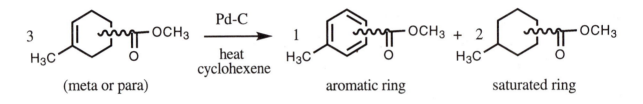

A. Preparation of Methyl (3 or 4)-Methyl-3-cyclohexenecarboxylate

Caution: Aluminum chloride is very hygroscopic, and releases HCl gas upon contact with water. Avoid breathing the dust or fumes, and keep all containers of AlCl₃ well-sealed when not in use.

Procedure

    Oven dry (preferably overnight) a 100 mL round bottom flask with stir bar, then seal it with a septum and allow it to cool to room temperature. Quickly transfer <u>about</u> 500 mg of anhydrous AlCl₃* and immediately reseal the flask. Add 10 mL of hexanes by syringe. Immerse the flask in a room temperature water bath over a magnetic stirring motor. While stirring to suspend the AlCl₃, add methyl acrylate (3.6 mL) by syringe. After stirring a few minutes, add isoprene (4.4 mL) by syringe over a period of 2 minutes. Let the mixture stir at room temperature for 1.5 hours. With stirring, cautiously add 10 mL of 1 M HCl. After 5 minutes, allow the layers to separate, and pipette away the aqueous phase. Dry the organic phase with anhydrous magnesium sulfate, followed by filtration into a 50 mL round bottom flask. Remove remaining solvent by rotary evaporation.

* Speed is more important than accuracy here, especially in humid climates. The instructor may provide pre-measured samples of AlCl₃ to expedite this step.

Purification by vacuum distillation: Place a stir bar in the flask containing the crude product, and set up an apparatus for simple vacuum distillation (Figure 7). Plan to collect all of the distillate in a single *tared* flask; it is not necessary to collect separate fractions. Be sure that the thermometer is placed properly to measure the distillation temperature, and use very good stirring to avoid the bumping to which vacuum distillations are prone. Apply full aspirator vacuum followed by heating as necessary to accomplish the distillation. The reported boiling point is 91-93 °C at 20 torr; measure or estimate (p. 14) the pressure in your apparatus and use the boiling point nomograph (Figure 10) to estimate the expected boiling point at the pressure you are using. It will probably be necessary to insulate the distillation adapter and the upper portion of the flask to accomplish the distillation. Collect all of the distillate in the tared flask and record the temperature range over which the product distills. Do not distill entirely to dryness; stop the process when perhaps $1/2$ mL of liquid remains. When the distillation is complete, remove the heat and release the vacuum. Determine the weight of product and prepare a sample for capillary GC analysis using the "paper clip" method (p. 45; use either CH_2Cl_2 or hexanes, and use a temperature program of 100 °C to 140 °C at 5 °C per minute). At your instructor's discretion, obtain ^1H and ^{13}C NMR spectra.

B. Reaction with Palladium

Caution: Cyclohexene has a strong irritating odor. Handle this material only in a hood.

Procedure

In a 25 mL round bottom flask, place 1 mL of your Diels-Alder product and 2 mL of cyclohexene. Add 60 mg of 10% palladium on activated carbon, install a reflux condenser on the flask, and heat the mixture to reflux under inert atmosphere. After 1 hour, cool the mixture and filter off the palladium-carbon catalyst through a pipette containing a cotton plug and 2 cm of Celite. Apply aspirator vacuum (but no heat) to the filtrate to remove cyclohexene (*Caution: odor!*) until approximately constant weight is obtained. Prepare a sample for capillary GC analysis using the "paper clip" method and use the same temperature program given in part A. At your instructor's discretion, obtain ^1H and ^{13}C NMR spectra and analyze the aromatic region to assign the structure of the original Diels-Alder product. Explain the details of your analysis in your notebook writeup.

Questions

1. Draw the structure of the expected complex between methyl acrylate and AlCl$_3$.

2. You might have expected the ^{1}H NMR of the Diels-Alder products to show two methyl singlets for each isomer. Is this the case? Explain.

3. Why did the capillary GC chromatogram of the mixture after heating with palladium show three large peaks instead of two? Can you assign the peaks? (see discussion of predicting GC retention times)

Experiment 7: The Wittig Reaction

The Wittig reaction was the first general method for converting a ketone to an alkene with well-defined position of the carbon-carbon double bond. Because alkenes are themselves very useful in organic synthesis, the development of this method eventually resulted in a Nobel prize, which Georg Wittig shared with H. C. Brown (who developed the hydroboration reaction) in 1979. Although several more recent variations on the basic reaction have been developed using different phosphorus reagents, the original reaction remains a powerful method for the synthesis of alkenes. In the reaction below, we will be preparing an exocyclic alkene (i.e., C=C extending out of a ring). Exocyclic alkenes are generally less stable than the corresponding endocyclic alkenes (i.e., C=C inside a ring). Synthesis of exocyclic alkenes is possible only because the Wittig reaction does not allow equilibration between alkene isomers (unlike most acid-catalyzed alkene preparations).

The procedure here involves a two-step synthesis, shown below. First, benzyltriphenylphosphonium bromide is prepared by reaction of triphenylphosphine with benzyl

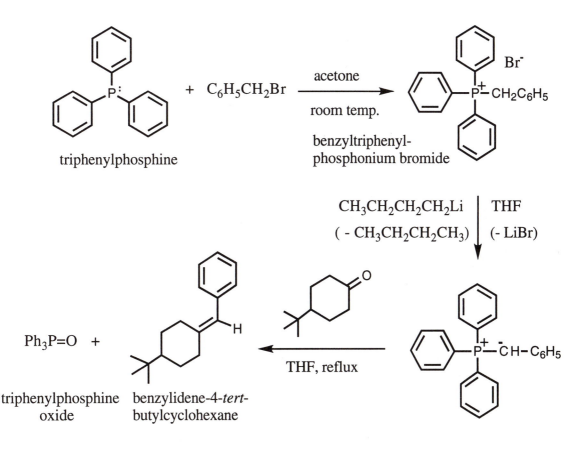

bromide. Although triphenylphosphine is only weakly nucleophilic, benzyl bromide is very reactive and the reaction is relatively rapid. In the second step, the phosphonium salt is deprotonated with butyllithium to make an *ylide* (a species with opposite charges on adjacent atoms), which in this case is a distinct red-orange color. Reaction of the ylide with a ketone proceeds by nucleophilic attack at the carbonyl group followed by elimination of triphenylphosphine oxide (which precipitates during the reaction).

A.   Preparation of Benzyltriphenylphosphonium Bromide

Caution: Benzyl bromide is a powerful lachrymator! Handle it and the crude reaction mixture only in a hood, and wear gloves.

Procedure

In a 50 mL Erlenmeyer flask, dissolve 5.0 g of triphenylphosphine in 20 mL of acetone. Add 2.40 mL of benzyl bromide by syringe, and seal the flask with a septum. Swirl the contents to mix the reactants, then set it aside for at least 1 hour; leaving it overnight or longer is preferable.

Note: After use, rinse the syringe and needle with concd ammonium hydroxide solution by repeatedly pulling up and expelling about 10 mL of the NH_4OH solution in a beaker. Disassemble the syringe and let it dry in the hood. Rinse it with acetone in the hood before cleaning.

Isolate the crystals by vacuum filtration, washing once with 10 mL of acetone. Transfer the crystalline solid to a *tared* round bottom flask, briefly dry the product under aspirator vacuum and record the weight of the phosphonium salt. Then apply aspirator or pump vacuum and heat the flask in an oil bath to about 110-130 °C for 30 minutes. After cooling and carefully removing any oil residues from the flask, record the net weight of product. At your instructor's discretion, obtain proton, carbon and/or phosphorus NMR spectra on the product. Approximately 50-60 mg can be dissolved in 0.4-0.5 mL of $CDCl_3$, though the mixture may require several minutes of warming and shaking.

B. Preparation of Benzylidene 4-*tert*-butylcyclohexane.

Caution: Butyllithium is potentially pyrophoric. Wear gloves and rinse the syringe and needle with hexanes immediately after use.

Procedure

In an oven-dried 100 mL round bottom flask with a stir bar place 3.00 g of benzyltriphenylphosphonium bromide. Then equip the flask with a condenser, and (preferably) put the apparatus under passive inert atmosphere by applying vacuum followed by inert gas using a three-way stopcock (see Figure 2). Suspend the phosphonium salt in 20 mL of dry tetrahydrofuran and add 6.9 mmol of butyllithium in hexanes (Note 1) over 1-2 minutes. Then weigh 4-*tert*-butylcyclohexanone (1.18 g) into a small test tube, and add this rapidly in one portion to the deep red reaction mixture by removing the condenser briefly. Reflux the reaction mixture for one hour, then allow it to cool to room temperature, followed by dilution with 20 mL of hexanes. Then filter the mixture into a round bottom flask, and concentrate the cloudy filtrate as much as possible by rotary evaporation, providing a mixture of liquid and solid. Dilute a small amount of the liquid for capillary GC using the "paper clip" method (p. 45); use a temperature program of 140 °C to 280 °C at 10 °C per minute. Purify the liquid portion by column chromatography on 20 g of silica gel using 100% hexanes. After TLC analysis, combine those fractions which contain only the strongly UV-active product (i.e., no ketone) and concentrate by rotary evaporation, followed by application of aspirator vacuum to achieve constant weight. At your instructor's discretion, obtain a capillary GC and ^{1}H and ^{13}C NMR spectra on the product.

Notes

1. The volume used will depend on the concentration of the reagent as provided by your instructor. A yellow-orange coloration should develop immediately upon addition of the first few drops of butyllithium; record how many drops are required in your reaction before the color persists. Some solid may remain even after butyllithium addition is complete.

Questions

1. Why do you suppose you were directed to rinse the benzyl bromide syringe with concd. ammonium hydroxide?

2. Based on the weight of your phosphonium salt before and after, why do you suppose heating was necessary? How could you verify your conclusion?

3. If you obtained or were provided with the proton-decoupled ^{31}P NMR spectrum of the phosphonium salt, account for the two sets of small satellite peaks at the base of the phosphorus signal. You should be able to verify your analysis using the ^{13}C NMR spectrum.

4. What was the limiting reagent in the formation of the alkene? Speculate on why the reaction was set up this way, and why heating was used (which is not necessary for most Wittig reactions).

Exploring Further...

1. At your instructor's discretion, you can determine why procedure B calls for heating. This is done by cautiously adding an equimolar amount (based on phosphonium salt) of acetic anhydride shortly after the ketone addition, and omitting the reflux entirely. Partition the reaction mixture between water and ether, dry over magnesium sulfate and analyze by GC-MS.

2. At your instructor's discretion, carry out the Wittig reaction using a different phosphonium salt and compare the results, particularly the capillary GC chromatograms of the crude products, with those obtained for other phosphonium salts. Use the procedure above, but remember to use the same molar amount (not the same weight) of phosphonium salt and add the ketone to the ylide as a *solution* in THF *slowly* over a few minutes. Heating should not be necessary.

3. The chirality of benzylidene 4-*tert*-butylcyclohexane can be illustrated using the Sharpless asymmetric dihydroxylation. One enantiomer is reported to react 9.7 times more rapidly than the other with AD-mix β (see: K. B. Sharpless, et al. *J. Am. Chem. Soc.* **1993**, *115*, 7864), so if the reaction is stopped prior to completion, the remaining alkene will be optically active. Using the procedure given in Experiment 10, carry the reaction to about 60% conversion using about 0.75 g of commercial AD-mix β per mmol of alkene. (In kinetic resolutions of this type, AD-mix β is more effective than the α version.) After workup, isolate the remaining alkene by chromatography and determine its specific rotation. Optically pure alkene is reported to have $[\alpha]_D^{25} = 33.3°$ (c = 1.12 , ethanol); see: S. Hanessian, et al. *Tetrahedron Lett.* **1992**, *33*, 7655.

Experiment 8: Preparation of 4,6,8-Trimethylazulene

The azulenes are an interesting class of aromatic hydrocarbons. This is partly because they are intensely colored (typically blue or purple) and partly because their reactivity is dominated by the molecule's ability to retain aromatic subunits when undergoing both electrophilic and nucleophilic attack. The parent hydrocarbon azulene, with its numbering system, is shown on the next page.

Azulenes were first noticed during the isolation of certain essential oils from plants. This was because some sesquiterpenes (15-carbon compounds derived from isoprene units) present in the essential oils would undergo isomerization to substituted azulenes during distillation or other manipulations, resulting in a distinctive blue or purple coloration. Some of these naturally occurring azulenes are shown here. Azulenes occur sometimes in petroleum, again coloring a certain portion of the distillate blue. It was not until the 1950s that synthetic procedures for the preparation of azulenes were developed. Guaiazulene is the only azulene which is commercially available at moderately low cost.

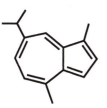

guaiazulene (blue)
from dehydrogenation
of patchouli and eucalyptus oils

chamazulene (blue)
from camomile, yarrow,
and wormwood oils

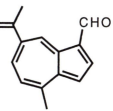

lactaroviolin (purple)
found in some
edible fungi

The reactivity of azulenes illustrates several features of nonbenzenoid aromaticity. The ring system contains 10 π electrons, and therefore qualifies as a $4n + 2$ aromatic system ($n = 2$). However, azulene is not as aromatic as the corresponding benzenoid system, naphthalene. Electrophilic attack always occurs at the 1 or 3 position, which results in formation of an aromatic cycloheptatrienyl cation. This intermediate can become entirely aromatic by loss of H$^+$, as benzene derivatives typically do after electrophilic attack. Conversely, nucleophilic attack always occurs at the 4, 6 or 8 position, resulting in the formation of an aromatic cyclopentadienyl anion. This anionic intermediate cannot easily become entirely aromatic again (why?).

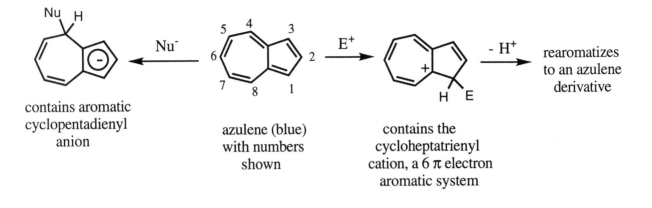

contains aromatic
cyclopentadienyl
anion

azulene (blue)
with numbers
shown

contains the
cycloheptatrienyl
cation, a 6 π electron
aromatic system

rearomatizes
to an azulene
derivative

In this experiment you will prepare 4,6,8-trimethylazulene by a multistep route involving two other interesting nonbenzenoid aromatics, a pyrilium salt and the cyclopentadienyl anion. Trimethylazulene has an intense purple color and is less volatile and easier to handle than azulene itself.

The pyrilium ion is an unusually stable oxonium ion (a carbocation α to an ether oxygen). Although carbocations are typically very short-lived and react with any conceivable nucleophile, the pyrilium salt you will be preparing is stable to the boiling alcohol solvent from which it is recrystallized! This stability is due to the fact that the ring is aromatic, having 6 π electrons just as benzene does. In order to isolate these salts, it is necessary for the anionic counterion to be non-nucleophilic. The well-known non-nucleophilic anions are tetrafluoroborate (BF_4^-), perchlorate (ClO_4^-) and hexafluorophosphate (PF_6^-). The perchlorate anion is potentially a powerful oxidant and can form shock-sensitive, explosive mixtures with organic compounds (which act as reducing agents). Although the published preparation of 4,6,8-trimethylazulene (K. Hafner, H. Kaiser *Organic Syntheses*, Coll. Vol. 5, Wiley, 1973, p. 1088) called for the use of the perchlorate salt, this experiment uses the tetrafluoroborate salt (A. T. Balaban, A. J. Boulton *Organic Syntheses*, Coll. Vol. 5, Wiley, 1973, p. 1112) to avoid any danger of explosion.

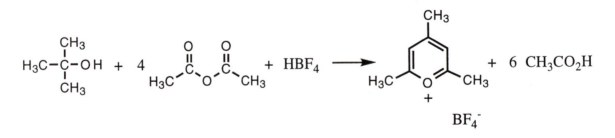

The reaction to form trimethylazulene, with a probable mechanism, is shown.

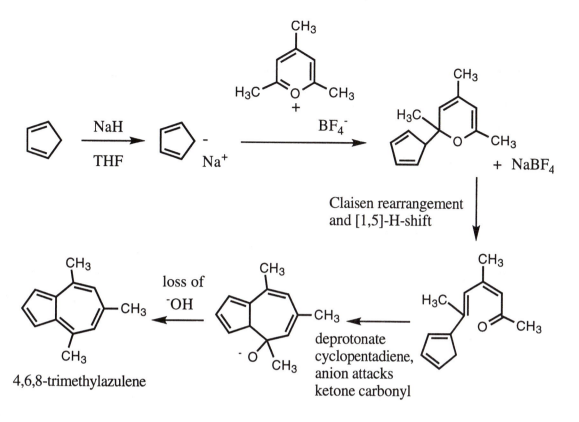

Like nearly all carbanions, the cyclopentadienyl anion, although aromatic, is destroyed by water or oxygen. Therefore, you will run this reaction (but not the isolation) under inert atmosphere as much as possible. The purification consists of (a) column chromatography followed by (b) extraction of the azulene into aqueous sulfuric acid and recovery by dilution, partial neutralization and re-extraction into hexanes. The solubility of azulenes in aqueous sulfuric acid is a special case of their reactivity with electrophiles, specifically reaction with H^+ as shown below. Most impurities are not soluble in aqueous sulfuric acid and are easily removed this way.

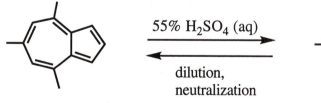

neutral: soluble in hexanes   cationic: soluble in aqueous H_2SO_4

A. The Preparation of 2,4,6-Trimethylpyrilium Tetrafluoroborate

Caution: Acetic anhydride is destroyed by water, so keep the flask covered with aluminum foil as much as possible. Avoid contact with acetic anhydride or fluoboric acid. Wear gloves, and wash with water immediately if contact occurs.

Procedure

Place a magnetic stir bar in a 250 mL Erlenmeyer flask, then add 50 mL of acetic anhydride and 6.0 mL of *tert*-butyl alcohol. Place a thermometer in the flask and seal the top of the flask with an aluminum foil collar. With good magnetic stirring, add fluoboric acid (48% in water; 8.2 mL, 63 mmol), **initially only a few drops at a time**, then in about 0.2 mL portions such that the final temperature reaches 90-100 °C. When the dark solution has cooled to about 60-80 °C, cool the flask in an ice-water bath; a crystalline precipitate should form. After cooling to 0 °C, add 100 mL of diethyl ether to complete the precipitation. After stirring at least an additional 10 minutes, the solid is filtered off and washed 1 or 2 times with small portions of ether.

Recrystallize the crude product from a 1:1 mixture of methanol and ethanol to which a few drops of fluoboric acid have been added. Use your 250 mL Erlenmeyer flask, and begin with 90 mL of solvent (plus a few drops of acid) heated to nearly boiling while stirring magnetically. Add additional solvent occasionally (perhaps 10 mL every 3 minutes or so) until the solid dissolves. Once the solid has dissolved, cover the top of the Erlenmeyer flask with aluminum foil and allow the solution to cool slowly to room temperature. When crystallization is complete, isolate the crystalline white solid by filtration and wash it sequentially with ethanol, ether and then hexanes. Transfer the product to a tared flask and dry under aspirator vacuum to constant weight.

B. The Preparation of 4,6,8-Trimethylazulene

Caution: Sodium hydride is quickly destroyed by atmospheric moisture; keep all vessels containing NaH tightly sealed as much as possible. Clean up any and all spills promptly. Use extreme caution when handling sulfuric acid. Wash immediately with water any acid allowed to spill or contact skin.

Preferably ahead of time, prepare 2-3 mL of cyclopentadiene as directed in Experiment 5A. If this cannot be used during the same laboratory period, it may be stored in a freezer overnight or in dry ice indefinitely.

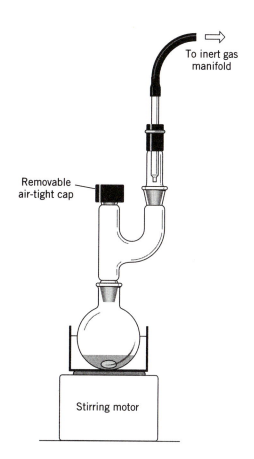

To inert gas
manifold

Removable
air-tight cap

Stirring motor

Figure 20: Reaction setup for 4,6,8-trimethylazulene preparation.

Procedure

Oven-dry (preferably overnight) a 100 mL round bottom flask with stir bar, distillation adapter, vacuum take-off adapter and 50 mL flask. While still warm, assemble these as shown (Figure 20) and let the apparatus cool under a flow of nitrogen. When cool, seal the top of the distillation adapter with a septum. Remove the 100 mL flask and quickly transfer 510 ± 30 mg of sodium hydride (Note 1). *Immediately tightly reseal both the flask and the NaH bottle.* Reattach the flask to the apparatus. Syringe 10 mL of dry tetrahydrofuran (THF) through the septum into the 100 mL flask. Surround the flask with a bath of cool water (15-20 °C) and suspend the NaH with magnetic stirring. Then syringe 1.4 mL of cyclopentadiene directly into the flask (not onto the walls of the distillation adapter). Gas evolution (H_2) should occur and a reddish coloration develop as cyclopentadienylsodium forms, which should be complete within 5 minutes. The water bath is then brought to approximately 40 °C. Over the next 30 minutes, 3.00 g of 2,4,6-trimethylpyrilium tetrafluoroborate is added by very briefly removing the septum as each small spatula-full is added, increasing the inert gas flow each time to prevent the entrance of air. Use your spatula to dislodge any appreciable amount of solid which accumulates in the distillation adapter. Keep the water bath at 40-45 °C during this time. When the addition is complete, increase the bath temperature to 80 °C and distill off the THF. When no more liquid will distill, add 10 mL of hexanes and distill this off as well.

This is a good stopping point. Seal the flask well and store it in your drawer.

Prepare a column for chromatography using 30 g (60 mL) of silica gel, slurry-packed with hexanes. Add enough hexanes to your crude product to give 4-5 mL of liquid, and suspend the brown residue as well as possible (a spatula helps). Transfer the slurry to a 15 mL polypropylene centrifuge tube and centrifuge the mixture to settle out the brown solid, then carefully apply the purple supernatant to the column. Remember to rinse the original flask and the residue in the centrifuge tube with 1 or 2 small portions of hexanes, followed by centrifugation, to get most of the azulene onto the column. Elute the column with hexanes, collecting the rapidly moving purple band in a 250 mL round bottom flask. When you have collected the entire purple band, take a sample for GC by diluting 3-5 drops in 1 mL of hexanes, and analyze this using a temperature program of 140 °C to 250 °C at 10 °C per minute. Then concentrate the solution by rotary evaporation to a volume of approximately 50 mL, and transfer this to your separatory funnel. Add 10 mL of 55% (w/w) aqueous H_2SO_4, cap the funnel securely and invert it. Carefully release any pressure which may have developed, then reclose the stopcock and shake the funnel vigorously for a total of 2 minutes, releasing the pressure as necessary. Allow the layers to separate completely, then drain off the lower aqueous layer into a 250 mL Erlenmeyer flask. If the hexanes layer retains a substantial purple coloration, repeat the extraction with one more portion of sulfuric acid. Then treat the acid layer (or combined acid layers) with 30 mL of water and 15 mL of hexanes. An intense purple coloration should appear immediately. Fully neutralize the acid solution by **cautiously** treating it with 3 g of NaOH in 15 mL of water for every 10 mL portion of acid you used (i.e., if you used 20 mL of acid, use 6 g of NaOH). Meanwhile, clean your separatory funnel. Transfer the water/hexanes mixture to your separatory funnel and drain off the aqueous layer. Pour the intensely purple hexanes layer into a 125 mL Erlenmeyer flask, and re-extract the aqueous layer with 15 mL portions of hexanes until the aqueous phase is no longer purple. Dry the combined hexanes extracts over magnesium sulfate. At this point, prepare a sample for GC by diluting 2 or 3 drops in 1 mL of hexanes, and analyze as before. Filter the solution into a tared 250 mL flask and concentrate by rotary evaporation, followed by application of aspirator vacuum until constant weight is achieved. At your instructor's discretion, obtain ^1H and ^{13}C NMR spectra, GC-MS and a UV-visible spectrum (use ethanol solvent; see instructor for details) Submit your product in a labeled vial to your instructor. Your notebook writeup should include a calculation of the percent yield of the azulene, and interpretation of the NMR, MS and/or UV spectra.

Notes

1. The weight of sodium hydride given here refers to dry (oil free) material. If the sodium hydride is used as a dispersion in mineral oil you must (a) weigh out more to account for the weight of the oil, and (b) you should wash the oil away by rinsing 2 or 3 times with hexanes (see instructor for details).

Questions

1. What was the concentration of the fluoboric acid that you used?

2. What color changes did you notice during the treatment with 55% H_2SO_4?

3. In the context of NMR, how many different types of hydrogens are in the azulene product? How many types of carbons?

4. What changes, if any, would you expect in the ^1H NMR spectrum if trimethylazulene in $CDCl_3$ were treated with D_2O and a small amount of D_2SO_4? Any changes in the ^{13}C NMR?

5. Which carbon(s) in the trimethylpyrilium salt came from the tert-butyl alcohol and which came from the acetic anhydride? Can you give a mechanism for the reaction?

6. Does trimethylazulene absorb more strongly in the visible or in the UV region of the spectrum? What color does the λ_{max} correspond to?

Exploring Further...

With your instructor's permission, consider the following experiments:

1. Azulenes react immediately with trifluoroacetic anhydride at room temperature. Choose conditions (solvent, concentrations), and isolate the product by column chromatography.

2. Dissolve 20-40 mg of trimethylazulene in 0.5 mL $CDCl_3$ and treat this with 0.2 mL of 20% D_2SO_4 in D_2O. Shake the tube occasionally and monitor by proton NMR over several days.

Experiment 9: The Asymmetric Synthesis/Resolution of Ibuprofen

Analgesics are compounds which relieve pain without causing unconsciousness. How these compounds work is not always understood, but they are thought either to act directly on the nervous system and/or to block the formation of prostaglandins. There are many compounds which function as analgesics, most notably aspirin (acetylsalicylic acid) and acetaminophen (Tylenol™). Most analgesics also function as antipyretics (fever-reducing drugs).

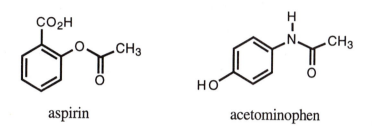

aspirin   acetominophen

One class of analgesics are the 2-arylpropionic acids, with the general structure shown below. These compounds have some advantages over aspirin and acetaminophen. The arylpropionic acids are several times stronger analgesics, and the dose/response curve covers a greater range. For example, taking twice (or four times) as much of an arylpropionic acid produces twice (or four times) the response. For aspirin or acetaminophen, the maximum effective dose occurs at much lower levels and taking more may *not* result in increased response. The arylpropionic acids have the antipyretic properties common to analgesics, but also anti-inflammatory action (reduces swelling and inflammation). This is accomplished by interfering with the production of prostaglandins which would otherwise cause this reaction. Interestingly enough, the nature of the aryl group is directly related to the potency of the drug, with some being at least ten times stronger than others. Ibuprofen (below) is the most common example, and until recently was the only 2-arylpropionic acid available without prescription.

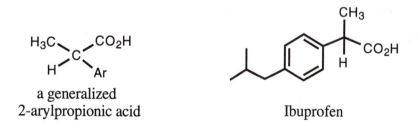

a generalized
2-arylpropionic acid   Ibuprofen

Unlike aspirin and acetaminophen, the arylpropionic acids are chiral and can exist in either of two enantiomeric forms. Because many of the materials of which biological systems are made are chiral and consist of only one mirror image, each enantiomer of an arylpropionic acid exhibits different biological effects. For this reason, some of these analgesics are sold as a single mirror image (the biologically active one). However, preparing these materials in one enantiomeric form is more expensive than making the racemic mixture, so if the inactive enantiomer does not cause unwanted side effects, a drug company may sell the product in racemic form (with FDA approval). Also, in at least some cases, conversion of the inactive enantiomer into the biologically active one (chemically referred to as an *inversion*) is known to occur in the body. The net result is that some of these materials are sold racemic and some as single mirror images. In the case of ibuprofen, only racemic versions are currently available. However, the desired activity is known to reside in only the (S) isomer, although the (R) isomer is converted to the (S) isomer in the body. There is an advantage to using the (S) isomer: it is reported to be effective within 12 minutes, while the racemic mixture requires 36 minutes. A great deal of research is being done in an effort to provide ibuprofen and other chiral drugs in nonracemic forms cheaply enough to be practical (see *Chemical & Engineering News*, October 9, 1995, pp. 44-74).

There are two basic ways to obtain nonracemic materials:

(a) Resolution: It is usually possible to separate the enantiomers of a racemic material. This is typically done by conversion of the compound to diastereomeric derivatives, followed by separation of the diastereomers (usually by recrystallization) and regeneration of the original compound. This process is referred to as a resolution. The drawbacks are that the process is time-consuming (costly) and that you lose at least half of the material as the unwanted enantiomer.

(b) Asymmetric synthesis: In this approach, you attempt to prepare the compound in nonracemic form directly from a nonchiral precursor. Because nonchiral reactants always give racemic products, this process requires the intervention of a chiral and nonracemic reagent or catalyst. The advantages are that the process gives a higher yield of the desired enantiomer and has the potential to be easier than a classical resolution. The success of an asymmetric synthesis is determined by the relative amounts of each enantiomer which are formed, and this is usually expressed in percent enantiomeric excess (% ee). The enantiomeric excess of a sample may be calculated using Equation 12, and can be considered a measure of how much of a pure enantiomer is present in addition to the racemate.

$$\% \text{ ee} = \frac{(\% \text{ major enantiomer} - \% \text{ minor enantiomer})}{(\% \text{ major enantiomer} + \% \text{ minor enantiomer})} \qquad \text{(Equation 12)}$$

An interesting asymmetric synthesis of (S)-ibuprofen was reported in 1989 (R. D. Larsen, et al. *J. Am. Chem. Soc.* **1989**, *111*, 7650-51). In this approach, racemic ibuprofen was converted to a nonchiral ketene by means of an acid chloride. Treatment of the ketene with a chiral and nonracemic alcohol produces a mixture of esters where the new chiral center is predominantly of the (S) configuration. Hydrolysis of the ester then provides (S)-ibuprofen in relatively high yield.

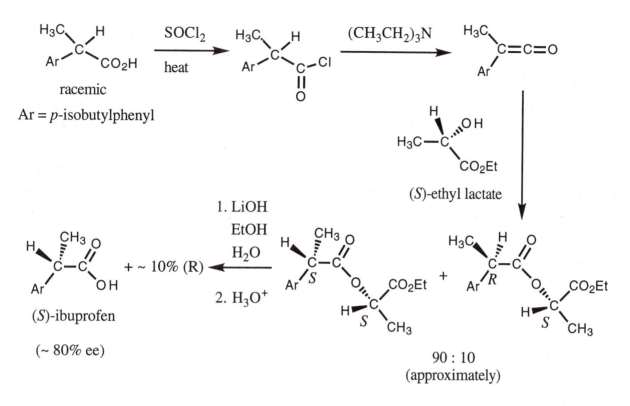

Ar = *p*-isobutylphenyl

(S)-ethyl lactate

(S)-ibuprofen

(~ 80% ee)

90 : 10
(approximately)

Procedure

A. Isolation of Racemic Ibuprofen

Determine and record the weight of six ibuprofen caplets. Powder the material in a mortar and pestle and transfer the powder to a centrifuge tube. Add 7 mL of hexanes, cap the tube securely and warm the contents in a water bath to 65-70 °C. Venting any pressure as necessary, shake the tube intermittently for about 5 minutes, keeping the contents warm in the bath. Then immediately centrifuge the tube briefly (max speed about 30 seconds) to obtain a clear layer. Pipette the clear supernatant directly into a 50 mL flask. It is important to transfer only the clear supernatant; transfer the liquid through a filter pipette if necessary. Also, if you wait too long, the ibuprofen may crystallize (especially on cold days). Dilute one drop of this solution to 1 mL with hexanes in a GC vial for capillary GC analysis; use a temperature program of 160 °C to 220 °C at 5 °C per minute. Then concentrate the remaining solution by rotary evaporation (Note 1). Complete the drying with your hood aspirator to constant weight, and record the net weight of your product (Note 2). At your instructor's discretion, obtain ^{1}H and ^{13}C NMR spectra. Be sure the GC chromatogram (reduced to 64%) is labeled with the temperature program and taped into your notebook.

Your material must be dry (no water present) for the next step. If in doubt, see your instructor.

Notes

1. The ibuprofen solution is prone to bumping! Be sure you are starting with a clean rotovap, and leave it clean for the next person.

2. If you obtain less than 700 mg of ibuprofen, re-extract the original solid once more with 6 mL of hexanes. (A good yield is 950 mg.)

B. Conversion of Ibuprofen to the Acid Chloride

Briefly oven-dry a 25 mL flask containing a 1" (medium) stir bar. Meanwhile, prepare a calcium chloride drying tube using a polypropylene centrifuge tube (bore a small hole at the bottom to attach a 6" length of $1/_8$" plastic tubing terminating in an 18 gauge needle, all connections sealed with Parafilm: see Figure 21). Cap the hot flask with a septum and allow it to cool. Then weigh into the flask 500 mg of the ibuprofen you isolated last lab period. Add one drop of *N,N*-dimethylformamide (DMF) and recap the flask with the septum. Insert the needle end of the drying tube assembly through the septum, then add 0.5 mL of thionyl chloride by syringe while stirring. Gas will be rapidly evolved, so be sure the drying tube is open at the far end! Using an oil bath on a hot plate, heat the mixture to 80 °C and hold it there for about 15 minutes. Then cool the flask to

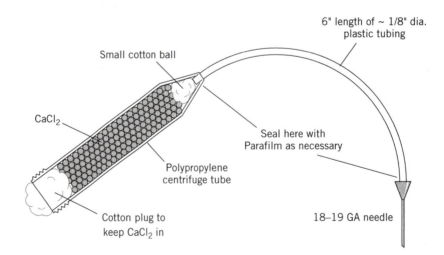

Figure 21: Drying tube assembly for thionyl chloride reaction.

room temperature. Remove excess thionyl chloride under aspirator vacuum and immediately put the acid chloride under inert atmosphere using a three-way stopcock (see Figure 2).

If you do not have time to go on to the next part (60 minutes minimum), you may stop here. Your acid chloride MUST be well sealed; use Parafilm on the outside of the stopper. Replace the stopper with a septum (quickly) before proceeding to the next step.

C. Conversion to the (S)-(+)-Ethyl Lactate Ester *via* the Ketene

Dissolve your acid chloride in 5 mL of dry hexanes. With magnetic stirring, add 1.0 mL of dry triethylamine* by syringe. Allow the mixture to stir for 1 hour, during which time it will become increasingly yellow and a thick precipitate will form. Then add 5 mL more hexanes and cool the mixture to 0 °C. Add 0.275 mL of (-)-ethyl lactate dropwise (wait 5 sec between drops); however, if the yellow color should fade entirely at any point, stop the addition and note how much ester remained. Make sure the solution comes in contact with any yellow material on the walls of the flask.

* Triethylamine is conveniently dried by shaking 10 mL of the amine with 2 g of freshly powdered sodium hydroxide in a centrifuge tube, followed by centrifugation. The resulting supernatant should be clear and colorless.

If you are nearing the end of the lab period, you may stop here.

Isolation of the ibuprofen ethyl lactate esters: After the yellow color of the ketene has subsided, add 10 mL of water and stir to dissolve the solid. Carefully pipette away the aqueous (lower) layer and wash the organic phase with 1 M HCl (5 mL) followed by 10% sodium bicarbonate (5 mL). Dry the organic phase with magnesium sulfate. Transfer the solution to a 50 mL round bottom flask through a filter pipette, rinsing the magnesium sulfate once or twice with hexanes. Dilute 1 drop to 1 mL of dichloromethane for GC analysis; use a temperature program of 160 °C to 220 °C at 5 °C per minute. Then concentrate the material by rotary evaporation and bring the flask to constant weight using aspirator vacuum.

D. Conversion to (S)-(+)-Ibuprofen

Add a stir bar, 2 mL of ethanol and 2 mL of water to the crude ester. Then add LiOH•H_2O (28 mg per 100 mg of crude ester) and let the mixture stir at room temperature for 30-45 minutes. At this point, the ester should no longer be evident by TLC (Note 1). Using 2 mL of water and 3 mL of hexanes to rinse, transfer the mixture to a separatory funnel or a 10 mL vial. Use a pipette to remove the organic (upper) phase, then re-extract the aqueous phase once more with hexanes (3 mL) and three times with ether (3 mL each). Keep in mind that it is the aqueous phase which contains the desired product; the organic extracts may be discarded once you have completed the isolation. Then acidify the aqueous phase to pH ~ 2 using 3 M HCl, and extract the solution twice with ether (5 mL each). Dry the combined extracts over magnesium sulfate and filter the solution into a tared round bottom flask using a pipette containing a small cotton ball. Concentrate the solution by rotary evaporation and purify the ibuprofen by column chromatography on silica gel (10-12 g) using 20% ethyl acetate in hexanes. The pure ibuprofen may be isolated as a liquid rather than as a solid, and the chemical purity can be verified by capillary GC as in part A. Determine the enantiomeric purity by either polarimetry (Note 2) or by conversion to the (+)-methylmandelate esters (Experiment 11C).

Notes

1. Develop the plate in 30% ethyl acetate in hexanes, and stain with PMA. There may be a UV-active spot at the same R_f as the ester, but it will not stain with PMA as the ester does.

2. Enantiomerically pure ibuprofen has been reported to have $[\alpha]_D^{25} = 59°$ (c = 2, ethanol).

Remember that the concentration c is in units of g per 100 mL, so $c = 2$ refers to a solution of 20 mg per mL.

Questions

1. Do you think the ibuprofen was present in the tablet as the acid or as a salt? Explain how your observations allow you to determine this.

2. When the ketene was reacted with (-)-menthol, the ester product gave a single peak by capillary GC and a single spot on TLC. The ^1H NMR spectrum of the (-)-menthyl ester after purification by column chromatography (chemical purity > 99%) is provided. With respect to asymmetric induction, what was the approximate outcome of this experiment? (Hint: Compare the CH_3-CH-CO_2- methyl signals at 1.5 ppm to what you would expect.)

Exploring Further...

1. You can quickly screen a wide variety of chiral alcohols and amines for the degree of asymmetric induction using capillary GC. In small vials, place about 5 mg of a chiral alcohol (or 1° or 2° amine) to be tested, and add by syringe 25 µL of the ketene solution (which is ~ 0.5 M). Because there is excess alcohol (or amine), the yellow color should fade entirely. Add 1-2 mL of hexanes, and analyze by capillary GC using the following program: initial temp = 160 °C; rate = 5 °C per minute; final temp = 250 °C; final time depends on the size of the ester; esters from "small" alcohols (like ethyl lactate) will come off before the final temperature is reached, while "large" alcohols give esters which may require several minutes at the upper temperature. Diastereomers are generally recognizable because (a) their peaks are large compared to those of impurities present, (b) they generally give two peaks very close together, and (c) the ratio of diastereomers will probably be less than 3 or 4 to 1 (usually not very high asymmetric induction). However, be aware that not all diastereomeric esters will be separated even by capillary GC, so a single large peak probably indicates a lack of separation rather than extremely high asymmetric induction.

¹H NMR of ibuprofen (-)-menthyl ester (360 MHz).

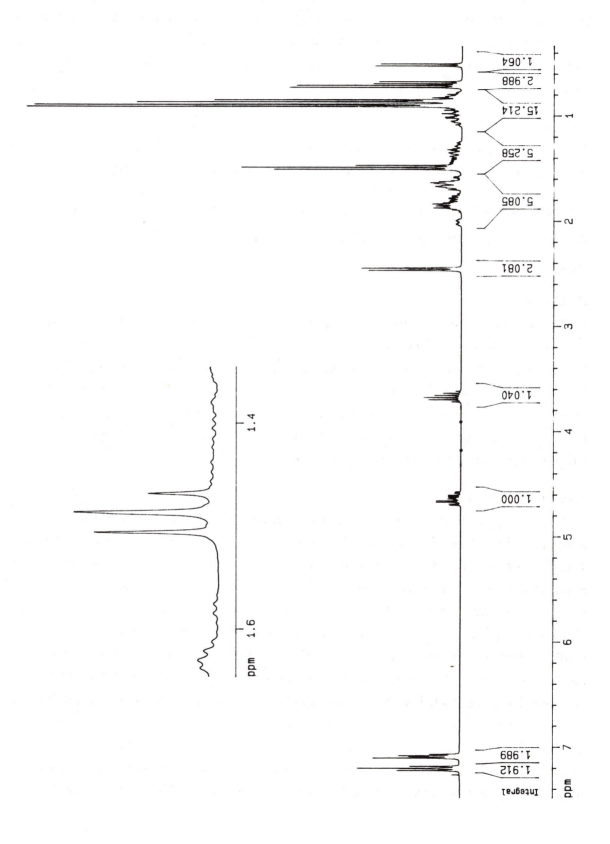

Experiment 10: The Sharpless Asymmetric Dihydroxylation

Asymmetric syntheses can be classified into two basic types, those which are stoichiometric in the asymmetric material and those which use only small amounts of an asymmetric catalyst. Catalytic asymmetric processes can have two major advantages over stoichiometric methods: (a) less of a possibly expensive chiral reagent is required and (b) because the process is catalytic, there is no need to remove covalently-attached residues in a separate step (cf., Experiment 9).

One of the most efficient catalytic asymmetric methods is the Sharpless asymmetric dihydroxylation reaction (see K. B. Sharpless, et al. *Chem. Rev.* **1994**, *94*, 2483; and K. B. Sharpless, Z.-M. Wang *J. Org. Chem.* **1994**, *59*, 8302). The reaction is an asymmetric variant of the dihydroxylation of alkenes with osmium tetroxide, with two important differences: (1) a chiral amine complexes the osmium to create an asymmetric environment, and (2) only catalytic (small) amounts of osmium are required because a stoichiometric oxidant ($K_3Fe(CN)_6$) is present to re-oxidize the Os^{VI} intermediate to Os^{VIII}. Osmium tetroxide is a volatile solid, extremely toxic (especially dangerous to the eyes) and very expensive, so the development of dihydroxylations which were catalytic in osmium represented a significant advance. Mixtures of K_2OsO_4 (which is oxidized to Os^{VIII}) with the re-oxidant and the chiral amine ligand are sold commercially under the trade name AD-mix. Amine ligands (shown below) have been developed which yield either enantiomer of the diol product, and so two versions are available: AD-mix α and AD-mix β, which give opposite enantiomers as the major product. A typical reaction using *trans*-stilbene is shown below.

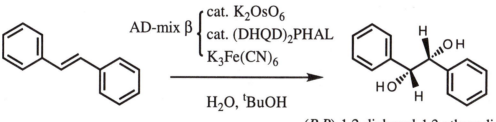

(R,R)-1,2-diphenyl-1,2-ethanediol

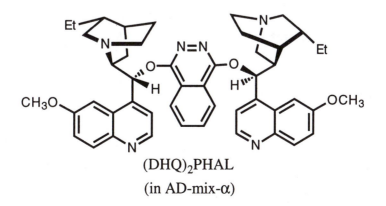

(DHQ)$_2$PHAL

(in AD-mix-α)

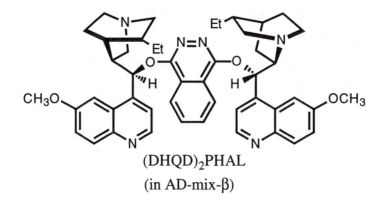

(DHQD)$_2$PHAL

(in AD-mix-β)

There are at least two methods which may be used to determine the enantiomeric purity of the 1,2-diol product. The optical rotation may be measured and compared to that reported for the enantiomerically pure material. Alternatively, the diol may be reacted with the chiral derivatizing agent (-)-menthyldichlorophosphate, as shown. This produces two possible diastereomers which (in theory) have different properties. In practice, the diastereomers are best distinguished using ^{31}P NMR.

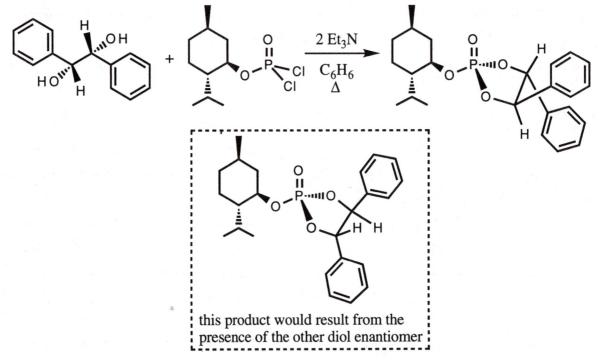

this product would result from the presence of the other diol enantiomer

A. Preparation of 1,2-Diphenyl-1,2-ethanediol

Procedure

To a 25 mL round bottom flask containing a magnetic stir bar is added AD-mix α or β (1.12 g), 5 mL of deionized water, and 5 mL of *tert*-butyl alcohol. Stir the two-phase mixture vigorously while adding *trans*-stilbene (180 mg), and continue the stirring for 1-2 days (all at room temperature). Then add sodium sulfite (1.5 g) and continue stirring for 45 minutes. Transfer the mixture to a separatory funnel, rinsing the original flask with ethyl acetate (a few mL), and allow the layers to separate. Remove the organic phase and extract the aqueous phase once more with ethyl acetate (5 mL). Dry the combined organic phases over magnesium sulfate, then filter into a tared flask and concentrate by rotary evaporation to constant weight. This should yield a white solid. At your instructor's discretion, prepare a sample for capillary GC (use a temperature program of 80 °C to 240 °C at 10 °C per minute), obtain ^{1}H and ^{13}C NMR, and proceed to determine the enantiomeric purity by one of the two methods below.

B. Enantiomeric purity by optical rotation
Procedure

If the purity of your product is low (< 95%), it may be recrystallized from a *small* amount of toluene. Determine from your instructor what volume of solvent is required to fill the available polarimeter cell. Then use a volumetric flask to prepare a solution in ethanol containing 25 mg of diol per milliliter. Depending on the volume of the cell, this may require combining your product with that of other students. Measure the optical rotation at 589 nm (sodium D line) and compare it to the published value ($[\alpha]^D = 94$ ° (c = 2.5, EtOH)).

C. Enantiomeric purity by derivatization/^{31}P NMR

Derivatization. In a dry septum-capped vial containing a stir bar, combine 25 mg of your diol, 0.8 mL of dry benzene (Note 1), 41 µL of dry triethylamine (Note 2) and 40 µL of menthyldichlorophosphate and heat at 60 °C for 30 minutes. Cool the mixture and transfer the contents to an NMR tube through a filter pipette, rinsing with a little dry benzene only if necessary to have sufficient volume for the NMR instrument. Obtain the proton-decoupled ^{31}P NMR spectrum between ± 20 ppm of the phosphoric acid reference. All peaks are observed as singlets in the proton decoupled spectrum. However, if proton decoupling is not used, the couplings to ^{1}H which would be observed are given here in parentheses. The diastereomers are observed at 13.65 and 13.95 ppm (d, J = 7.6 Hz), while any residual menthyldichlorophosphate gives a signal at 6.2 ppm (d, J = 9.3) and a small amount of dimenthylchlorophosphate impurity in the derivatizing reagent appears at 3.9 ppm (t, J = 8.3).

Notes

1. Benzene is easily dried by passage through dry alumina. If the available NMR instrumentation requires a deuterium lock, C_6D_6 may be used as part or all of the solvent. If the magnet homogeneity is good, it is often possible to run the sample unlocked.

2. Triethylamine is conveniently dried by shaking 10 mL of the amine with 2 g of freshly powdered sodium hydroxide in a centrifuge tube, followed by centrifugation. The resulting supernatant should be clear and colorless.

Questions

1. What is the stereochemical relationship between (DHQ)₂PHAL and (DHQD)₂PHAL? (Hint: The reported specific rotations for the enantiomerically pure materials are +336 and -262˚ (both c = 1.2, methanol), respectively).

2. Would you expect the Sharpless asymmetric dihydroxylation to give higher, lower or the same enantiomeric purities with cis-stilbene? (draw the expected diol)

Experiment 11: Identification and Optical Purity Analysis of a Commercial Arylpropionic Acid Analgesic.

In addition to ibuprofen, there are several 2-arylpropionic acid analgesic pharmaceuticals currently in use. In this experiment, you will extract such an analgesic from a pharmaceutical preparation, determine its structure using IR, NMR and mass spectroscopy, and finally determine if it is racemic or not by GC analysis of diastereomeric esters.

To determine whether or not a compound is racemic, we must have some way of distinguishing between enantiomers. Unfortunately, this is rather difficult because enantiomers have identical properties under most conditions. For example, (*R*) and (*S*) enantiomers (and any mixture of the two) will give identical GC and TLC chromatograms, NMR and mass spectra, etc. Only in the presence of other chiral and non-racemic materials will the two enantiomers behave differently (and even then not necessarily *very* differently). Examples of this would be how different mirror images of drugs have different biological effects, or how enantiomers can sometimes be separated by chromatography *if* the stationary phase is chiral and nonracemic. Another way to analyze enantiomers is to convert them to diastereomers by reaction with an appropriate chiral and optically pure compound. For example, a mixture of (*R*) and (*S*) carboxylic acids esterified with a pure (*S*) alcohol would yield a mixture of (*R,S*) and (*S,S*) diastereomers, which (in theory) have different properties (NMR spectra, capillary GC retention times, etc.). In practice, such diastereomers often have extremely similar properties and, for example, may not be separable chromatographically (though the differences are more often observable in high-field proton NMR spectra).

A convenient and mild method for the esterification of carboxylic acids involves treatment of a mixture of the acid and alcohol with dicyclohexylcarbodiimide (DCC). The DCC activates the carboxylic acid toward nucleophilic attack (i.e., converts the acid OH group into a good leaving group), and in the process is converted to dicyclohexylurea (DCU), which is evident as a precipitate as the reaction proceeds. The reaction is catalyzed by 4-(dimethylamino)pyridine (DMAP). For the mechanism of this reaction, see J. March *Advanced Organic Chemistry*; Wiley, 1992, p. 395.

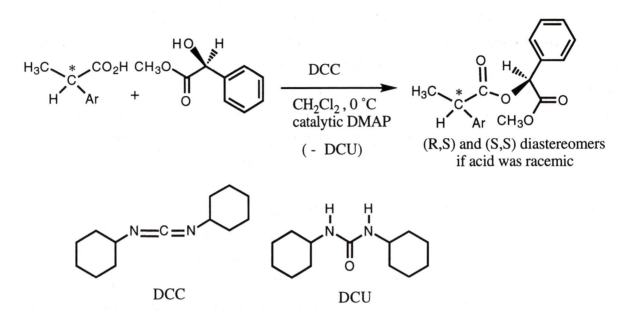

In this experiment, you will:

(a) Isolate an unknown arylpropionic acid from a commercial analgesic preparation, and determine the structure using NMR, IR and GC-mass spectroscopy. Because carboxylic acids often respond poorly to chromatography, the GC-MS will be done on the methyl ester rather than on the free acid.

(b) Determine if the compound is racemic or not by conversion to (*S*)-(+)-methyl mandelate esters followed by capillary GC analysis.

There will be 200 mg of active ingredient in the preparation you are given.

Procedure
A. Isolation of the Arylpropionic Acid

Record the code letter of your unknown and describe its appearance in your notebook. Determine the weight of the analgesic preparation, transfer the solid to a centrifuge tube and add 6 mL of water. Acidify to a pH of (approximately) 2 by adding 1 M HCl in 1 drop portions followed by shaking and testing with pH paper. Record in your notebook how many drops were required. Then extract twice with 4-mL portions of dichloromethane, removing the organic (lower) phase each time by pipette (centrifuge as necessary to get a clear layer). Dry the combined organic extracts with anhydrous magnesium sulfate and filter into a tared round bottom flask. Then concentrate the solution by rotary evaporation. Complete the drying under aspirator vacuum to constant weight, and record the net weight of your product. At your instructor's discretion, obtain melting point (for solids only), IR, and proton and carbon NMR spectra. Most of the analgesics are readily soluble in $CDCl_3$, but in one case you will be unable to dissolve more than about 50-60 mg in 0.4-0.5 mL of this solvent. For GC-MS (and possibly IR), first convert a small amount of the acid to its methyl ester as described below. Use this data to determine as much of the structure as possible, showing your reasoning in your notebook. Be sure all data, chromatograms and spectra are reduced appropriately and taped into your notebook.

B. Conversion to methyl ester for GC-MS analysis

Weigh 15-20 mg of your carboxylic acid into a test tube (approx. 10 x 70 mm). Add 0.5 mL of methanol and one drop of concd sulfuric acid. While shaking the tube side-to-side, warm the reaction mixture in a water bath at 60-70 °C for two minutes. Let the mixture cool and cautiously add 2 mL of 10% aqueous sodium bicarbonate and 2 mL of hexanes. Mix the two phases well with a spatula, then allow the layers to separate. Carefully remove the hexanes (upper) phase (see Figure 17) and pass it through a filter pipette containing 1-2 cm of granular sodium sulfate, collecting the filtrate in a vial. Analyze the compound by GC-MS, using a program of 160 °C to 260 °C at 5 °C per minute. Remember that the methyl ester will have a molecular weight 14 amu higher than the original carboxylic acid. Because most of the methyl esters are liquids, it may be simpler to obtain the IR spectrum on the ester. In addition, some of the IR signals for the methyl esters tend to be narrower than for the free acids. The IR sample can be prepared by evaporating most of the hexanes off, then allowing the last several drops to evaporate on a sodium chloride plate. If your ester is a solid rather than a liquid (i.e., it crystallizes on the salt plate), you will have to obtain the IR spectrum using one of the solid-sample techniques (KBr pellet or mineral oil mull).

C. Determination of Enantiomeric Composition.

Caution: DCC is a sensitizer and can cause allergic reactions in some people on subsequent exposure. Handle DCC only in the hood and wear gloves when syringing it.

Esterification with (+)-methyl mandelate: Into a dry test tube with a small magnetic stir bar, weigh approximately 20 mg of your carboxylic acid and 20 mg of (+)-methyl mandelate. Dissolve these into 1 mL of dry dichloromethane, add a tiny crystal of DMAP and seal the test tube with a rubber septum. With stirring, cool the solution in an ice bath for a few minutes, then add (by 1 mL syringe) 120 μL of 1 M DCC in CH_2Cl_2 provided by the instructor (Note 1). A white precipitate should form within 30 seconds. Allow the solution to warm to room temperature and stir for 45 minutes. At your instructor's discretion, you may monitor the completion of the reaction by TLC. Then add 50 μL of water to the reaction and let it stir another 5 minutes. Filter the solution into a round bottom flask through a pipette containing 1 cm of Celite™, rinsing the flask and column with 1-2 mL of dichloromethane. Dilute 5-8 drops of this to 1 mL with CH_2Cl_2 for GC analysis (240 °C at 2 °C per minute to 280 °C, hold at 280 °C for 4 minutes). In your notebook, detail any evidence of diastereomers which may be apparent from this analysis.

Notes

1. Diisopropylcarbodiimide may be substituted for DCC. The former is a liquid which may be syringed neat. However, little or no precipitate will form if diisopropylcarbodiimide is used.

Questions

1. What percent of the analgesic you were given was active ingredient? (You should have two answers: one based on the weight you were told would be present, and one based on the weight you actually obtained.)

2. Do you think your analgesic was present as the acid or as a salt? Explain. (Hint: Estimate the amount of acid you had to add.)

3. In cases where a single peak is observed for the methyl mandelate ester, how could you distinguish whether a single enantiomer was present or if the diastereomers simply were not separated?

4. If for any reason the esterification reaction had not gone to completion, could this have influenced the ratio of ester diastereomers or not? (Does each enantiomer of the acid have to react at the same rate with the alcohol, or can they react at different rates?)

5. In DCC-promoted esterifications, it is common for some (usually only a small amount) of the carboxylic acid to be converted to an N-acyl urea, of general structure shown below. Show the most plausible mechanism by which this might be formed in the reaction.

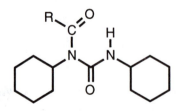

an *N*-acyl urea

Experiment 12:   Unknown Reaction Product Analysis.

In many real-life situations, you will know what materials are involved in a reaction but will not necessarily recognize immediately what the product(s) are. In the following experiment you will encounter one such reaction, and you are to determine the structure(s) of the major product(s) using the instrumental techniques we have learned in this course (though you may **not** use the library search feature of the GC-MS). However, you are not limited to these techniques and may employ, for example, qualitative organic analysis methods if you choose to. Expect to isolate one major product unless directed otherwise. Some details are left to your expertise, such as how to do column chromatography (and what solvents to use). Also, you may deviate from the procedures given, *but you are responsible for the outcome.* It is a good idea to obtain capillary GC chromatograms on both the product <u>and</u> the starting material so that (a) you can verify that the reaction proceeded at least mostly to completion, and (b) the retention times can be compared to give you some idea of the "size" of the product relative to the starting material. Be sure you include all aspects of the usual notebook report, including the reaction (structures) drawn out, a complete table of reagents, a full discussion of your interpretation of the data by which you determined the structure(s) of the reaction product(s) and a proposed mechanism by which the reaction occurred.

Your instructor will provide you with an experimental procedure prior to the start of this experiment.

Index

NOTES

NOTES

NOTES

NOTES

NOTES

NOTES

NOTES